When Telephones Reach the Village

The Role of Telecommunications in Rural Development

COMMUNICATION AND INFORMATION SCIENCE

A series of monographs, treatises, and texts
Edited by
MELVIN J. VOIGT
University of California, San Diego

William C. Adams • Television Coverage of the Middle East
William C. Adams • Television Coverage of International Affairs
William C. Adams • Television Coverage of the 1980 Presidential Campaign
Alan Baughcum and Gerald Faulhaber • Telecommunications Access and Public Policy
Mary B. Cassata and Thomas Skill • Life on Daytime Television
Hewitt D. Crane • The New Social Marketplace
Rhonda J. Crane • The Politics of International Standards
Herbert S. Dordick, Helen G. Bradley, and Burt Nanus • The Emerging Network Marketplace
Glen Fisher • American Communication in a Global Society
Oscar H. Gandy, Jr. • Beyond Agenda Setting
Oscar H. Gandy, Jr., Paul Espinosa, and Janusz A. Ordover • Proceedings from the Tenth Annual Telecommunications Policy Research Conference
Edmund Glenn • Man and Mankind: Conflict and Communication Between Cultures
Gerald Goldhaber, Harry S. Dennis III, Gary M. Richetto, and Osmo A. Wiio • Information Strategies
Bradley S. Greenberg • Life on Television: Content Analyses of U.S. TV Drama
Bradley S. Greenberg, Michael Burgoon, Judee K. Burgoon, and Felipe Korzenny • Mexican Americans and the Mass Media
Cees J. Hamelink • Finance and Information: A Study of Converging Interests
Heather Hudson • When Telephones Reach the Village
Robert M. Landau, James H. Bair, and Jean Siegman • Emerging Office Systems
James Larson • Television's Window on the World
John Lawrence • The Electronic Scholar
John S. Lawrence and Bernard M. Timberg • Fair Use and Free Inquiry
Robert G. Meadow • Politics as Communication
William H. Melody, Liora R. Salter, and Paul Heyer • Culture, Communication, and Dependency:
Vincent Mosco • Broadcasting in the United States
Vincent Mosco • Policy Research in Telecommunications: Proceedings from the Eleventh Annual Telecommunications Policy Research Conference
Vincent Mosco • Pushbutton Fantasies
Kaarle Nordenstreng • The Mass Media Declaration of UNESCO
Kaarle Nordenstreng and Herbert I. Schiller • National Sovereignty and International Communication
Harry J. Otway and Malcolm Peltu • New Office Technology
Ithiel de Sola Pool • Forecasting the Telephone
Dan Schiller • Telematics and Government
Herbert I. Schiller • Information and the Crisis Economy
Herbert I. Schiller • Who Knows: Information in the Age of the Fortune 500
Jorge A. Schnitman • Film Industries in Latin America
Indu B. Singh • Telecommunications in the Year 2000
Jennifer Daryl Slack • Communication Technologies and Society
Dallas W. Smythe • Dependency Road
Sari Thomas • Studies in Mass Media and Technology, Volumes 1–3
Barry Truax • Acoustic Communication
Georgette Wang and Wimal Dissanayake • Continuity and Change in Communication Systems
Janet Wasko • Movies and Money

In Preparation:

William Dutton and Kenneth Kraemer • Modeling as Negotiating
Fred Fejes • Imperialism, Media, and the Good Neighbor
Howard H. Fredericks • Cuban-American Radio Wars
Kenneth Mackenzie • Organizational Design
Armand Mattelart and Hector Schmucler • Communication and Information Technologies
Everett Rogers and Francis Balle • The Media Revolution in America and in Western Europe
Keith R. Stamm • Newspaper Use and Community Ties
Robert S. Taylor • Value-Added Processes in Information Systems
Tran Van Dinh • Independence, Liberation, Revolution

When Telephones Reach the Village

The Role of Telecommunications in Rural Development

Heather E. Hudson
University of Texas at Austin

Ablex Publishing Corporation
Norwood, New Jersey 07648

Library of Congress Cataloging in Publication Data

Hudson, Heather E.
  When telephones reach the village.

  (Communication and information science)
  Bibliography: p.
  Includes indexes.
  1. Telecommunication—Developing countries. I. Title.
II. Series.
TK5102.5.H83 1984   384′.09173′4   84-18409
ISBN 0-89391-207-7

Ablex Publishing Corporation
355 Chestnut Street
Norwood, New Jersey 07648

Contents

To the memory of Dr. Martha Wilson, who worked tirelessly to bring the benefits of satellite communications to Alaska natives.

Acknowledgements

Many people made major contributions to the material which appears in this book. Douglas Goldschmidt, Andrew P. Hardy, and Edwin P. Parker co-authored with me the literature review prepared for the International Telecommunications Union (ITU) entitled *The Role of Telecommunications in Socio-Economic Development: A Review of the Literature with Guidelines for Further Investigation*. Support for the literature review and the field studies in Alaska and the South Pacific by the author was provided by the ITU, under a research project headed by William B. Pierce, Jr. Support for the preparation of a synthesis of 15 ITU-funded case studies was provided by the Development Centre of the Organization for Economic Co-operation and Development (OECD), under a project directed by Nicolas Jéquier, whose project summaries were used in the preparation of chapters 3, 4, and 5. Other major supporters of my field research have been the U.S. Agency for International Development, the World Bank, the National Library of Medicine, and the Canadian Department of Communications. Valuable insights and encouragement have come from many sources including Helen Barr, Clifford Block, Richard Butler, Dennis Foote, Lyndsay Green, Michel Guité, Meheroo Jussawalla, Gerry Kenney, Patsy Layne, Lyle Nelson, Emile McAnany, Bella Mody, Jim Morris, Donna Pace, Walter Parker, Theda Pittman, Robert Saunders, Wilbur Schramm, Douglas Ward, Bjorn Wellenius, and Florence Woolner. Several colleagues from the Canadian Department of Communications may claim responsibility for getting me interested in telecommunications for remote areas. And friends and colleagues across the north and throughout the developing world taught me about communications, but also about development. Credit for this book should be shared with these mentors. The responsibility for the accuracy of the contents and the views expressed is mine alone.

Preface

My first experience with rural two-way communications came in the form of crackling high frequency (HF) radios in the Canadian north. I had been sent by the Canadian Department of Communications to investigate complaints from residents of remote villages on the Labrador coast about their poor communications services. I accompanied an engineer who was to determine the technical causes of the problems. I was to investigate the need for communications—why people wanted communications, what kind of information they needed, and where and with whom they needed to communicate.

I learned from that short field project about the importance of two-way communications for isolated villagers—as an emergency lifeline, a "gossip" network, and an organizational tool to coordinate their travels, to check on shipping schedules, and to keep in touch with relatives scattered among the many coastal settlements.

I heard about the importance of two-way communications again and again as I investigated the requirements for communications among native people in Canada's remote north. Two-way communication—between isolated communities and with major regional centers—was always the first priority. Broadcasting—in the form of radio and television—was also requested, but was clearly seen as secondary to reliable two-way communication. Indeed, a group of chiefs from remote northern Ontario surprised government officials by citing reliable and accessible two-way communications as their only priority. When asked about other items on their list (which usually included better housing, electrification, improved airstrips, etc.), they responded: "We need communications first. Once we get that, we'll discuss the other things we need."

Requests such as these led to a northern pilot project sponsored by the Canadian Department of Communications to learn more about native communications needs through demonstration projects which involved primarily two-way HF

radio networks and low-powered community radio broadcasting stations. It occurred to me during my evaluation of these projects that I had not learned about
the role of two-way communications in graduate school; at that time the focus
was on the use of mass media for development, primarily for educational applications. Why had two-way communications been overlooked, I wondered.

Other fieldwork and research projects reinforced the importance of interactive
networks. In Alaska, I worked on the evaluation of biomedical communication
projects using NASA's ATS-1 and ATS-6 experimental satellites for communication between village health aides and physicians at regional hospitals. Later,
in planning communication projects in several developing countries for the U.S.
Agency for International Development and the World Bank, I was again struck
by the importance rural people and those providing services to rural people
attached to reliable two-way communications. In seeking to understand more
about the role of interactive communications in development, I was drawn in two
directions—into applications of telecommunications for development through
project planning and evaluation, and into research on the role of telecommunications, particularly in rural development.

This book attempts to summarize what we know about the role of telecommunications in the development process, and to raise questions which we still
need to answer.

Development and Telecommunication: A Theoretical Framework

Chapter 1

Introduction

1.1   The Problem

This book examines the role of telecommunications in the development process. In doing so, it attempts to shed light on a relationship that seems both obvious and obscure.

To the user in a rural or isolated area, the importance of telecommunications is self-evident. An Indian chief on a remote reserve in northern Canada needs to communicate with government officials about the village school. A village health aide in Alaska needs to communicate with a doctor to get advice on treating a patient. An aboriginal elder in the Australian outback needs to communicate with a legal organization concerning land claims. A village trader in India needs to communicate with a wholesaler in the city to order merchandise. An agricultural extension agent in the Cook Islands needs to communicate with the shipping company to find out when the ship will pick up the citrus fruit crop from his island.

Despite the virtual self-evidence of these needs, the role that telecommunications plays in development has remained obscure. At the macro level, it seems obvious that telecommunications contributes to the efficient operation and productive growth of an economy. But does telecommunications contribute to development, or is the growth of telecommunications a manifestation or even a consequence of economic development? There is a strong correlation between a country's wealth (measured, for example, in Gross National Product per capita), and its investment in telecommunications (one indicator of which is telephone density, or telephones per 100 population). Yet wealthy countries have more available resources to invest in telecommunications, which can, in turn, contribute to the process of increasing their wealth. Poorer nations have many demands on limited resources, and may therefore assign telecommunications a relatively low priority—thus missing an opportunity to use telecommunications as a tool to further their development.

The relationship of telecommunications to development, although obvious, has remained obscure because of the difficulty of disentangling the chicken-and-egg relationship of telecommunications investment and economic growth. There have also been few attempts to isolate the impact of telecommunications in cases where new services have been introduced or their role in certain sectors can be identified and analyzed.

However, in the past few years, there has been a growing interest among researchers in examining the impact of telecommunications in both industrialized and developing societies. Among the organizations which have stimulated this interest through support of rural telecommunications projects and/or research are the International Telecommunications Union (ITU), the Development Centre of the Organization for Economic Cooperation and Development (OECD), the U.S. Agency for International Development (AID), the United Nations Educational, Scientific and Cultural Organization (UNESCO), the World Bank, and the Inter-American Development Bank (IDB).

The purpose of this book is to bring together the research in this field in order to make it more widely available, and to put the research questions and findings within a development framework. The book's emphasis is primarily on rural development because a majority of the population in the developing world lives in rural areas, and it is the capability of modern telecommunications to bridge distances more reliably than in the past that now makes investments in telecommunications technology for rural areas attractive. Yet, the question of benefit from this investment that would justify the cost remains, particularly for rural telecommunication services. While demand in urban areas generally is great enough that revenues are certain to exceed costs, in rural areas, the issue is often more complex. Revenues may not, at least in the short term, cover the capital and operating costs of the rural telecommunications system. Yet indirect benefits to users and to the rural economy as a whole may more than justify the costs. It is these indirect benefits which are the focus of this book.

Part I provides a theoretical framework for an analysis of the role of telecommunications in the development process, with particular emphasis on rural development. Part II reviews the literature and key recent studies under the following headings:

- macro level studies
- rural development and social services
- commerce and industry

Part III examines issues concerning the provision of telecommunications for rural development including appropriate research methodologies, planning considerations, and policy issues.

1.2. Telecommunications: An Ignored Medium

Although the technology of telecommunications is changing very rapidly, the basic means of using electromagnetic waves to convey information over distance have been understood and used for more than a century. The telegraph dates from the 1830s; it accompanied the railways that tied together the industrial and commercial centers in Europe and pushed back the frontiers in North America. In 1976, the world celebrated the centenary of the invention of the telephone. By that time, the telephone was virtually taken for granted in the industrialized world. There were even some towns and cities (such as Washington, D.C. and Palo Alto, California) that had more telephones than people. And yet, in many parts of the world there was—and is—no telephone for hundreds of miles. In cities and towns, only a single telephone may be available for thousands of people.

Despite our hundred years of experience with telecommunications, it is surprising how little we know about the effects of this technology. Western researchers seem to have invested more time and effort in examining the effects of the electronic mass media—radio and television—than the effects of the instrument on their desks which links them to other people and organizations. There may be several reasons for this emphasis on mass media, or point-to-multipoint communications, rather than on point-to-point or interactive telecommunications. The study of the effects of electronic mass media grew out of an existing discipline—the study of the mass medium of print. Social historians have studied the impact of the print medium over the centuries. Social scientists have become intrigued with the effects of the mass media. Sociologists and social psychologists have studied the effects of the media on various elements of society including the family, children, ethnic and linguistic minorities, etc. Political scientists and psychologists also became interested in the role of media—particularly film and later radio and television—in influencing attitudes and behavior.

Funding for mass media research has come from different sources, including broadcasters and advertisers who are interested in knowing how many and what type of people they are reaching, and with what impact. In the United States, much research on the impact of media has been funded by government agencies and foundations concerned with the potential impact of the powerful medium of television, including means by which television and other media could be harnessed to improve education and to increase learning opportunities for both children and adults. Similarly, internationally, UNESCO sought to harness the potential of the mass media to help meet the needs of the developing world—to train and educate, and to provide development information to such groups as farmers and mothers of young children. Thus, UNESCO has funded projects and research on the effects of the mass media in developing countries and on applications of mass media for reducing illiteracy, increasing agricultural productivity, and improving nutrition and sanitation.

In contrast, point-to-point telecommunications were largely ignored by social scientists. Even psychologists interested in interpersonal or organizational communication tended to overlook the role of a medium in facilitating or affecting these forms of communication. Telecommunications was left largely to engineers, many of whom worked on improving and advancing the technology because they intuitively saw its benefits. Some researchers with technical backgrounds such as Colin Cherry (1971) drew attention to the impact of the telephone, and conducted their own research on its effects. On an institutional level, technical organizations led the quest for a better understanding of the effects of telecommunications. In Britain, the Post Office has funded research and development on the use of telecommunications in business and the home; Canada's Department of Communications includes divisions concerned with social and economic planning and policy, as well as groups of technical managers and scientists responsible for technological research and management of the frequency spectrum. In the United States, the National Aeronautics and Space Administration (NASA) provided the opportunity for experimentation in applications of satellite technology for education and social services as part of its Applied Technology programs.

Also, criteria used to assess the benefits of telecommunications were based on direct return on investment. Telephone utilities, whether privately or publicly owned, were generally required to set rates that covered their costs and allowed for an additional fixed "rate of return." Much of the excess of revenues over costs was reinvested in expansion and upgrading of the telecommunications system; the rest became profit for private companies, or revenues that could be used to subsidize other sectors, such as postal services, for government-owned entities. As revenues generally easily exceeded costs by a comfortable margin, there was little incentive to look for benefits beyond those which turned up on the balance sheet. There were, however, implicit assumptions about the value of providing telephone service in rural areas, as rates were generally set so that revenues from the profitable urban and interurban services subsidized rural services. Some countries developed special programs to extend telecommunications to unprofitable areas, based on intuitive assumptions, if not concrete evidence, of the benefits of such services in rural areas. Perhaps the best known of such programs has been administered in the United States by the Rural Electrification Administration (REA) which is part of the Department of Agriculture.

1.3   Diversity in the Developing World

There is tremendous economic diversity among the nations which are known as developing countries or the "Third World." The World Bank clusters these countries into three groups:

- low-income nations (GNP per capita $400 or less);
- middle-income nations (GNP per capita up to $5100);
- capital-surplus oil exporters (World Bank, 1983).

However, there is often great variation in economic conditions within developing nations. Urban populations continue to expand at alarming rates, exacerbating problems of poor living conditions and insufficient jobs. But the majority of the poor of the developing world still live in the rural areas.

Developing regions and nations can be characterized by both poverty and diversity. Wellenius (1978) describes the diversity of living conditions and development problems in developing countries and refers to two types of poverty: absolute poverty (annual income $50 or less in 1969 dollars) and relative poverty (incomes less than one-third of the national average income of the country). Wellenius emphasizes the importance of rural development:

> Of the total 750 million poor in the developing countries, 600 million (80%) including virtually all the absolute poor, live in rural areas. 40% of all people in these countries' rural areas are poor. Improving the lot of rural areas is thus central to any development effort.

For these reasons, the primary focus of this book is on the role of telecommunications in rural development. In addition to economic benefits, social benefits are considered that are likely to contribute to improved quality of life in rural areas. These emphases reflect current concerns with equity, decentralization, and self-reliance. To achieve these goals, developing nations must pay greater attention to rural human development.

1.4   The Telecommunications Environment

The growth in telephone investment in the post–World War II era has been phenomenal: in 1945 there were 41 million phones worldwide; by 1982 there were more than 494 million, an increase of more than 1,200%. In 1982, the total expenditures for telecommunications worldwide amounted to more than $77 billion.

Pitroda (1976) points out that 80% of the world's telephones are in North America and Europe. The capitalist countries have by far the largest share, with 86% of the world's telephones for a total population of 759 million, while the second or socialist world has 7% for a population of 1.3 billion, and the third world has 7% for a population of 2 billion.

Wellenius (1977) notes that there are not only major differences in telephone penetration between developing countries and developed countries (e.g., 18%

vs. 27.1% in 1972) but among developing countries (e.g., Argentina has 3.1% telephones per 100 inhabitants vs. .03 in Upper Volta—a difference of 270 times).

In Latin America, Wellenius comments that there are less than 5 phones per 100 inhabitants, and since 63% of all phones are in three countries (Argentina, Brazil, and Mexico) the actual average for less developed Latin American countries is even smaller. Okundi (1978) states that there were only 1.4 telephone instruments per 100 inhabitants in Africa in 1977, or .7 per hundred in black Africa alone.

1.5   The Financial Environment

Chasia (1976) points out that the effects of foreign finance so far have been to "widen the cleavage between the rural and urban areas in the field of telecommunications." International development institutions have placed a low priority on telecommunications. Financial institutions have taken the position that telecommunications should generally be self-supporting in the short term. Loans have generally been approved only if there is a high likelihood of a healthy internal financial rate of return. In practice, this requirement has led to financial support primarily for installing and upgrading urban facilities and interurban trunk routes with less support for rural service.

A second factor is lack of knowledge of the role of telecommunications in development. Both national planners and international agencies have considered telecommunications an urban luxury for wealthy businessmen and matrons. It appears that, at least in the past, major multilateral and bilateral development agencies have not been convinced that telecommunications projects would benefit the rural poor.

World Bank policy has been that utilities should be financially viable and recover the full cost of service from their tariffs. Through 1980, telecommunications loans and credits accounted for approximately 3.6% of total lending. Saunders (1982) states that internal financial rates of return attributable to ten recent telecommunications projects approved for World Bank support range from 31% to 35%, averaging 18%. Recent World Bank policy has placed increased emphasis on rural components of telecommunications projects, in line with the Bank's basic human needs policy to reach the rural poor. The Bank encourages needs assessment and evaluation through the establishment of socioeconomic research units within the national telecommunications administrations. For examples of research conducted by these entities, see Kaul (1978) and Saunders (1978).

Through December 1977, the Inter-American Development Bank (IDB) had loaned $196.8 million for telecommunications projects, or only 1.6% of its total loans to date, ranking this sector next to last in sectors financed by the IDB.

Only recently has the IDB begun to finance rural telecommunications projects. In 1976, it loaned $29 million to Colombia toward the construction of public telephones in 2,200 rural communities. A loan of $12.2 million to Costa Rica in 1977 was for construction of 56 telephone exchanges and 1,300 public telephones located in rural communities. In 1978, Ecuador borrowed $9.6 million for construction of 128 telephone exchanges and 254 public telephones in rural communities (Gellerman, 1978).

During the past 20 years there have been many advances in communications technology that could benefit the developing world. In particular, communication satellites have proved to be an appropriate means of serving rural and remote areas. The INTELSAT organization began to provide international satellite services in 1965, and within a decade had established reliable links between developing countries and the rest of the world for the first time. However, experiments and pilot projects during the 1970's demonstrated that this technology could also be used for telecommunications and broadcasting services within developing countries. For more information on the potential of satellites to serve developing regions, see Chapter 7.

During the 1970s, the United States demonstrated the possibilities of using satellite technology for education and social services through its Applied Technology Satellites (ATS series) built by the National Aeronautics and Space Administration (NASA). Experiments with ATS-1 showed the potential of satellites to support rural health and education services in Alaska and the South Pacific. After experimentation at home, the U.S. loaned the ATS-6 satellite to India where it was used to deliver educational television to over 2400 villages, each equipped with satellite antennas and television sets made in India.

At TELECOM 75, the international telecommunications exhibition held in Geneva in 1975, a model of an air-transportable earth station, developed by the ITU in collaboration with the Federal Republic of Germany and the Symphonie Organisation, was demonstrated for the first time.

By 1980, the French/German Symphonie satellite had demonstrated the potential applications of satellite systems in several developing nations of Africa and Asia. In North America, the US/Canadian Communications Technology Satellite (CTS) and Canada's Anik-B satellite had given further evidence of the potential of satellite services, particularly for scattered users and remote regions. Operational (as opposed to experimental) satellite service was now a reality for more than one hundred villages in Alaska (using the RCA SATCOM satellite) and many remote settlements in northern Canada (using the ANIK system). Domestic satellites were beginning to proliferate in the U.S., and several European systems were planned.

TELECOM 79, held in conjunction with the 1979 World Administrative Radio Conference (WARC), exhibited the latest developments in satellite technology and their applications for rural telephony, broadcasting, and remote sensing—with particular emphasis on the requirements of developing nations. However, the design of commercial satellite systems has not been optimized for rural regions with low traffic volume. To help achieve its goal of assisting developing countries to gain access to appropriate and affordable communications facilities, the ITU has proposed a design for a satellite system specifically to meet the needs of rural and remote areas of developing regions. This concept is known as GLODOM (for Global Domestic Satellite System). For a description of the proposed GLODOM system, see Chapter 7.

1.6   The Policy Environment

The International Telecommunications Union (ITU) is a consortium of 155 member governments joined together through the International Telecommunications Convention. Now a member of the United Nations (UN) family of agencies, it dates from 1865 when European governments founded the International Telegraph Union to set standards and rates for telegraph transmissions across national borders. It became the International Telecommunications Union in 1934. At present about 25% of its members could be considered developed or industrialized nations, while 75% are developing nations.

The ITU mandate includes:

- Technical and procedural (administrative) regulations and planning provisions for the effective utilization of telecommunication technology and its appropriate integration and operation within the world telecommunication networks and systems, including planning and use of radio frequency spectrum and satellite orbits;
- Recommendations for the standardization of equipment, functions and performance to enable interconnection between systems and networks, as well as advice and guidance to member countries for the planning of services, and for production in the manufacturing sectors.

Contributions to standardization and advice are achieved through the Consultative Committees: the CCIR (International Radio Consultative Committee), and CCITT (International Telephone and Telegraph Consultative Committee). Technical, procedural, and planning provisions are deliberated in worldwide or regional conferences, the agreements of which create treaty obligations.

The recent (1979) World Administrative Radio Conference (WARC) placed particular emphasis on the role of telecommunications in development and on the need for developing nations to gain access to appropriately designed telecommunications technology, particularly for domestic and rural services.

At the 1979 WARC, developing countries' needs for technical assistance were manifested in a series of resolutions and recommendations. Of particular relevance to the issues discussed in this book are resolutions which call for:

- Promotion of telecommunications in rural development for education health, agriculture, and other activities important for social and economic progress;
- International cooperation and technical assistance in the field of space radio communication, with the aim of making available any means of technical assistance in space communications in LDCs;
- Transfer of technology, urging administrations of developing countries to establish policies to strengthen international cooperation activities which will achieve transfer of telecommunications technology. (World Administrative Radio Conference of 1979, *Final Acts*)

Many of the issues concerning access to communications technology for development are central to the agendas of specialized ITU World and regional conferences including:

- Broadcast Satellite Planning Conference for Region 2 (The Americas), in 1983;
- WARC for planning HF Bands allocated to the Broadcasting Service, in 1984 and 1986;
- WARC on the Use of the Geostationary Orbit and the Planning of Space Services, in 1985 and 1988.

Concerns about the imbalances in the flow of information between the developed and developing world, and within those spheres, have led to a series of steps and decisions by other international bodies. Similarly, Third World concerns about the need for greater opportunities for participation by the developing nations in the world economy are manifested in calls for a new international economic order. Reports by the MacBride Commission (*Many Voices, One World,* 1980) and the Brandt Commission (*North–South,* 1980) have focused new attention on the role of information in development and international understanding. Such issues as the need for better communications facilities within developing countries which now have reliable international links, the requirements for appropriate technologies and techniques of technology transfer, and the need for new tariffs for conferencing and for international news exchanges to facilitate more communication from the developing to the industrialized world (South–North) and within the developing world (South–South) have all served to stimulate and integrate interest in telecommunications for development.

Among the steps in a program of action toward a new international economic order called for by the United Nations members are:

- adaptation of technology for development;
- increased international economic cooperation for development;
- information transfer for development;

- a Third Development Decade, with special emphasis on integrated rural development.

UN Resolution 35/201 (1980) concerning questions relating to information called for:

- collaboration for establishment of a new world information and communication order;
- a new International Program for the Development of Communication (IPDC);
- growing recognition of the importance of telecommunication infrastructures in creating a new order.

The IPDC has already become a reality. Its origins date from the 1978 UNESCO General Conference, where there was an implicit agreement that, in return for some moderation in developing countries' rhetoric concerning the need to institute formal measures to redress the imbalance in the North–South flow of information, the industrialized world would assist poorer nations in developing their communications capabilities more rapidly.

By 1982, a 35 nation Governing Council had been established, which set out criteria for projects and made the initial selection of projects to be funded through the IPDC. Priority was given to projects for:

- national planning for communications development;
- creation of infrastructures, particularly using indigenous technologies;
- institutional arrangements to facilitate free flow and a more balanced exchange of news and culture;
- training in research, planning, management, technology, production, and dissemination;
- regional and international cooperation, especially among developing countries;
- enhancement of "development communications" in support of education, agriculture, health, and rural development;
- increased access to the latest technologies, such as satellites and data banks.

Of 54 initially proposed projects, 19 were approved for funding support, with the primary emphasis on development and exchange of regional information. Both multilateral and bilateral funding will be accepted, although funding pledged to date falls far short of the resources required to carry out all of the proposed projects (Block, 1982).

Other international activities focusing attention on the contribution of telecommunications to development have included:

- UN Conference on Science and Technology for Development (1979);
- UNISPACE 82 (the Second UN Conference on the Exploration and Peaceful Uses of Outer Space);
- World Communications Year (1983) mandated by the UN General Assembly, with the theme of development of communications infrastructure.

1.7 Emphasis on Telecommunications

In the broad sense, the communications sector includes both broadcasting (e.g., radio and television distribution) and point-to-point communications (primarily telephone, telegraph, and telex). This book emphasizes telecommunications in the narrow sense, i.e. point-to-point or interactive communications at a distance.

There are good reasons for including broadcasting in a study of the developmental effects of telecommunications. There are many technical, economic, and organizational links between broadcasting and interactive telecommunications. Often they share the same transmission facilities—satellites, microwave, coaxial cable, etc. In many countries they are administered or regulated by the same governmental entity. Furthermore, broadcasting may have some very visible and well documented benefits, for example in increasing agricultural production through extension service programs, or extending and improving education and training through educational radio and television. However, as was pointed out above, there has been considerable research on the impact of the mass media, but very little in interpersonal media. The focus of this book, therefore, is on telecommunications in the narrow sense, i.e., primarily on the telephone system. However, some services included here such as teleconferencing may be considered hybrid as they are both distributive and interactive.

The primary emphasis of this book is on the role of telecommunications in developing regions, and much of the research was conducted in developing countries. However, field research conducted in rural areas of industrialized countries, such as Alaska, the mid-western United States, northern Canada, and farming regions of eastern Europe and the USSR, is also included because of the insights these studies can provide on the role of telecommunications in rural development.

1.8 The Need for Research

There are several reasons for the apparent underinvestment in the telecommunications sector of developing regions. Saunders states two primary reasons for World Bank involvement in the sector: to influence or refocus investment, pricing, and telephone allocation policies so that overall government objectives for development can be more efficiently pursued; and to promote institution building and policy improvements within the telecommunications sector and to help implement more rational long-term planning. However, World Bank projects must show direct and demonstrable impact on the lowest 40% income group in a country, and telecommunications projects that meet these criteria must compete for Bank funds with projects for other sectors such as agriculture, nutrition, and water supply for which there are not likely to be other substitutes for Bank funds (Saunders, 1982).

Saunders cites additional reasons for the generally low rate of investment in telecommunications in developing countries:

- a lack of enumeration and quantification of the benefits of telecommunications investment relative to what is done in other sectors;
- a perception that telecommunications investments, while profitable in a financial sense, confer direct benefits only upon a relatively narrow—and privileged—portion of the population of a developing country;
- tariff policies which in the short run do not promote an efficient allocation of tariff resources;
- institutional and organizational problems both within and exogenous to the telecommunications operating entities;
- a lack of available foreign exchange to invest in imported telecommunications equipment;
- lack of technically qualified personnel.

The first two issues in this list are the primary focus of this book. These issues are related because it is only through more research on the impact of telecommunications in the development process that we will gain a better understanding of both the direct and indirect benefits of telecommunications investment.

Since this manuscript was prepared, two sources have been published which complement this book. *Telecommunications and Economic Development* by Saunders, Warford, and Wellenius (1983) reviews numerous macro- and microeconomic studies, and examines telecommunications financing, tariff policies, and sector management. *Telecommunications for Development*, a report by Pierce and Jéquier (1983), is a synthesis of the ITU-OECD project on the contribution of telecommunications to economic and social development.

Chapter 2

Steps Toward a Theory

2.1   A Shortage of Theory

As a developmental tool, telecommunications has been largely ignored by planners and theorists. It is generally grouped with public utilities and infrastructure, ranking far below roads, power supply, water, and sanitation as investment priorities. Yet telecommunications is a tool for the conveyance of information, and it is the lack of consideration of the role of information in development theory that is perhaps more surprising. As Lesser and Osberg (1981) point out, only recently have economists begun to consider information as a variable rather than an assumption in analyzing markets and consumer behavior. Yet, developing nations are at an obvious disadvantage in their ability to obtain information required for optimum decision making. Researchers, including Cruise O'Brien and Helleiner (1982) and Jussawalla (1982), have begun to study the significance of access of information in international trade and negotiations involving developing countries.

This type of analysis does not appear to have been extended to development decision making *within* a country, particularly concerning its rural sectors. Yet it appears obvious that lack of access to critical information can place the rural resident at a major disadvantage, and can impede the improvement of rural social and economic conditions.

Another problem that may have impeded planners from adequate consideration of telecommunications as an input for development is the difficulty in measuring its impact. Unlike investments in new seed or fertilizer, for example, where a measured input can yield a measurable increase in output, telecommunications benefits do not lend themselves to such quantification. Telecommunications is considered a public good because its benefits include externalities—benefits that accrue not only to the person who places and pays for the call, but to the person called, and to other parts of the society. For example, if a field nurse calls a doctor for advice, the patient benefits, as might others with similar symptoms. It a local merchant orders spare tractor parts by telephone, the farmer benefits by not losing time because of idle equipment. These examples point to the more general problem of identifying and assessing the role of timely information in the development process.

2.2   Development Theory and Telecommunications

The major approach of development theories during the 1960s has been termed the "Dominant Paradigm." This orientation has in the past guided many national and international development programs.

Rogers (1976) outlines the assumptions of the dominant paradigm. First, definitions of development centered upon the rate of economic growth. Gross national product, and gross national product per capita were used as major indicators of the level of economic development. Quality of life indicators were not considered important, as presumably higher levels of income and employment would inevitably lead to an improved quality of life. The distribution of wealth indicated by these economic measures was also often not taken explicitly into account—distribution issues presumably would be minimized through high levels of economic growth. Unfortunately, the assumption that a growing amount of wealth would "trickle down" to those most in need proved to be inaccurate. Rather, the benefits of economic growth fell disproportionately on the already better-off. The distributional gaps between the rich and the poor widened rather than narrowed.

A second assumption was that the patterns of economic development exhibited in Europe and North America formed the blueprint of development for other nations. To reach the prosperity of western industrialized countries was seen as the goal of development strategies. Agricultural and rural development were relatively neglected as industrialization and urbanization were seen as the major routes to development. Unfortunately, industrial growth proved inadequate to provide the income and employment growth necessary to improve the conditions of a growing urban and rural poor.

Rogers (1976) highlights the elements of what he sees as new trends in development. Development is now seen as involving the equality of distribution of socioeconomic benefits. Decentralization of project planning and implementation is stressed, particularly at the district and village level. Also, self-reliance and independence in development through the use of local resources are emphasized. Finally, it is now generally recognized that there is a wide disparity in the conditions and prospects of developing nations, and that western development patterns are not necessarily appropriate models for other nations.

The role of telecommunications in developing regions and countries within this new "development" framework is uncertain. The lack of definitive studies concerning how telecommunications may affect economic development and be integrated in development strategies has caused difficulties for national planners, telecommunications planners, and international lending agencies such as the development banks, in determining both investment and pricing policies in the telecommunications sector to make the best use of limited capital resources for promoting national development.

Telecommunications as a factor in socioeconomic development was almost completely overlooked by communication researchers during the 1960s. Works by communications scholars such as Schramm (1964), Lerner (1958), and Shapiro (1967) (see also Frey, 1973) focused on mass communications as a tool for development through modernization. This was true for theoretical approaches as well as development strategies. The diffusion research school (e.g., Rogers) stressed interpersonal communication, but through personal contact rather than through telecommunications.

However, since the late 1960s, there has been a growing body of research, primarily by economists and technologists concerned with the possibility of using communication technology as a means of fostering economic development. The various studies resulting from this interest may be roughly grouped into three categories—statistical studies which relate some measure of telecommunications development with national economic development, studies of a generally nonempirical nature which attempt to delineate the broad influences of telecommunications on development, and case studies of particular telecommunications projects.

The first group of studies, while often provocative, has failed to provide the type of evidence necessary for national planning and investment decisions, particularly given the large capital investments necessary for telecommunications development. Second, case studies of individual projects and applications of communications indicate potential indirect benefits of telecommunications, but generally are not designed to control for other factors and cannot safely be generalized. The third group of studies typically shows relationships between telecommunications and development indicators (e.g., GNP or GDP per capita) but does not answer the question of causality. (See e.g., International Telegraph and Telephone Consultative Committee, 1968, 1972; Gilling, 1975; Cruise O'Brien et al., 1977.) While telecommunications investment and economic development are interdependent, it is not possible to infer from these studies the degree to which telecommunications contributed to economic development or telecommunications investment contributed to economic growth.

However, some recent studies sponsored by the ITU and the OECD Development Center have shed considerable light on these relationships. The major findings of these studies are reported in the following chapters. An overview of all the studies is found in Pierce and Jéquier, 1983. The individual study reports as cited in the bibliography of this book may be obtained directly from the ITU.

2.3   Constraints on the Impact of Telecommunications

The general socioeconomic value of telecommunications has been recognized by many writers. For example, Dickenson (1977) states: "If trade is the lifeblood of the economy, then telecommunications can truly be regarded as the nervous system of both the economy and society."

Telecommunications engineers working in developing countries have intuitively recognized the developmental benefits of telecommunications. For example, Rizzoni (1976) sees telecommunications investments as stimulants to development. He refers to Honduras' 4-step rural telephony program to extend service to over 150 rural communities: "To Honduras, a country presently with a modest economy, the development of telecommunications and other infrastructure will provide a much needed impulse to the national economy." He also finds Colombia's rural telephony program "a capillary telecommunication infrastructure conducive to development of . . . production potential." Developing country telecommunications planners such as Mgaya (1978) of Tanzania see telecommunications as a means of closing the gap between the elite and the common people.

But many planners and researchers have not adequately considered the social/political/economic environment in a country that may determine how and to what extent telecommunications may support development. As Wellenius points out:

"To understand telecommunications as a means rather than as an end in itself calls for looking into the functions it may have in a social context" (quoted in Blackman, 1977). He adds that there are theoretical arguments and empirical evidence which support the proposition that telecommunications are a necessary condition for development: "to a considerable extent, the slow development of rural telecommunications arises from an insufficient awareness of the role telecommunications can play in supporting such aspects of economic development as rural programs in agriculture, government administration, health, commerce, and transportation" (Wellenius, 1977).

Some researchers have identified constraints which they believe may greatly influence the impact of communications. Two types of constraints have been emphasized concerning the potential impact of telecommunications. One stresses the socioeconomic and political context. The second identifies other development components that must be present for communication strategies to be effective.

In a book edited by McAnany (1980), it is pointed out that structural constraints largely determine the scope for socioeconomic development. The authors define structural constraints as "societal obstacles that restrict the opportunities of an important number of individuals to participate fully and equitably in the development process and in the sharing of benefits of a given social system." They state that communication processes cannot be seen in isolation from the societal arrangements under which they have developed and the restrictions through which they exert their influence. One of the major structural constraints the authors emphasized is the land tenure system in Latin America, where farmers in the regions studied live on small plots suitable only for subsistence agriculture, while large landowners control much of the agricultural sector. Given this constraint, it is extremely difficult to help farmers improve their output and efficiency of production, or to have greater impact in the market system. The authors are therefore pessimistic about the ability of communication (primarily exemplified through mass media messages and the activities of extension agents) to contribute to meaningful development.

Karunaratne (1982) points out that telecommunications investment in a country must be examined in terms of who is served, and in the socioeconomic context of the country. For example, in Papua New Guinea, the density of telephones is nearly 1.3. However, only 0.6% of the total indigenous population of Papua New Guinea are telephone subscribers, while over 30% of the expatriates have telephones. Karunaratne states that the telecommunication system serves the export-oriented enclave economy, a situation symptomatic of the urban/rural and international dualism in many LDC economies. It must be pointed out that even when socioeconomic conditions do not impose major constraints and communications access is relatively widespread, communication alone will not necessarily stimulate development. Other factors must be present. Hornik (1980) refers to this requirement as complementarity: "Communication technology works best as a complement—to a commitment to social change, to changing resources, to good instructional design, to other channels of communication, and to detailed knowledge about its users." Complementarity may also refer to the requirement for other forms of infrastructure, including transportation, effective organizations, availability of supplies and expertise.

We may conclude that structural constraints in a society may preclude major socioeconomic change to which communication could contribute; however, barring such major constraints, telecommunications infrastructure plus other complementary infrastructure (development activities, entrepreneurial activities, transportation, a social service delivery system) can together lead to more economic growth and more effective social service delivery than when either or both of the two basic conditions are absent.

2.4   Information: A Neglected Element in Development

To begin to theorize about the role of telecommunications in development, it is necessary first to consider the role of information. Telecommunications systems, after all, are simply means of transmitting information instantaneously over distance. They have value only if there is value in the information being transmitted.

Lesser and Osberg (1981) point out that orthodox economic theory has paid little attention to the role of information in the economy. It is generally assumed that prices convey adequate information to allow the economic system to function. However, "should markets be in disequilibrium, should market participants differ in access to information, should imperfect competition exist in some commodities or futures markets but not exist in others, then prices will not suffice as guides to production and consumption." The Bayesian approach emphasizes the value of information. It is hypothesized that individuals must make decisions based on their estimates of the probability of future events, but that new information enables them to revise their estimates of the future. The decision problem of the individual comes at three different levels of generality:

- first, the method of acquiring information, or the "channel" to be used;
- second, how much to use the channel, given the cost of acquiring and processing messages;
- finally, the issue of information as an unusual economic good—a public good. "Unlike a private good, one person's consumption of information does not reduce the quantity of information available to others for consumption. This joint consumption property may mean that the socially desirable quantity of information will fail to be generated if the provision of information is left entirely to the market. It also implies that the marginal cost of information is zero for all users after the first. The marginal social cost of information flows is then only the marginal cost of information delivery and processing which is wholly a function of the modes of information-transfer available. Where communications systems exhibit increasing returns to scale, the average cost of such transfers declines as capacity increases" (Lesser and Osberg, 1981).

These issues have several implications for an understanding of the role of telecommunications in development. In general, telecommunications can be expected to generate more rapid, cheaper information flows; hence, the availability of telecommunications may alter the optimal channels or method of acquiring information. The economic impact of the use of telecommunications channels will depend upon:

- the increase in information flows achieved by telecommunications, i.e., the net advantage over the next best alternative in speed and informational content of messages delivered;
- the extent to which new messages change economic agents' opinions of the probability of events (i.e., their social receptiveness to new information);

- the economic pay-off of the new strategies motivated by the new information, which is dependent on many factors including the availability of necessary inputs of markets for outputs (Lesser and Osberg, 1981).

Thus, people must perceive that information transmitted has value to them in order for there to be any pay-off to any form of communication. Small farmers who have no fertilizer or who cannot borrow money to buy fertilizer or new high-yield seed will benefit little from information about new agricultural practices. However, farmers with greater receptiveness to innovation and with access to the necessary inputs or to sources of capital may benefit greatly from such information. Similarly, if there is only one outlet for a product, timely information on prices may be of little value.

Information can be seen as a necessary element for the efficient functioning of an economy. The information content of price signals, by themselves, is however, generally inadequate as the sole source of information in an economy. Also, prices themselves are information which must be disseminated.

Thus, the necessity both to disseminate price information and to supplement or correct the market information provided by prices gives rise to the need for information-transfer modes, i.e., the means to convey price and nonprice information.

Telecommunications may enable economic agents to develop new "channels" of information or to use existing channels more effectively, at lower cost. Where an optimal level of telecommunications services is not available, information flows will be restricted and economic agents will turn to other more expensive/less effective "channels" or means of obtaining information. Two sorts of private costs are implied—the higher direct cost of information and the cost of poorer decisions made in the absence of timely or adequate information. These private costs are likely to be incurred throughout the economy, not just among telecommunications subscribers, since uninformed economic agents often "follow the leader" and imitate the actions of the more informed. Where the quality and amount of information even of the more informed declines (due to the inadequate communications systems), the decisions of the leaders are likely to be worse.

Information is also critical to organizational and cultural development. In order to organize any activity, whether it be community visits by field nurses, formation of a cooperative, or political meetings, leaders must share information. Similarly, to sustain or strengthen a culture, people need to share information and transmit their heritage to their children or others who have left the traditional cultural areas. Telecommunications technologies allow information to be transmitted to community groups or large numbers of people through the mass media; they also facilitate the organization of social and cultural as well as economic activities through interactive point-to-point media such as the telephone. Thus it is the role and significance of information that we must seek to understand in order to determine the impact of telecommunications in the development process.

2.5   Steps Toward a Theory: Major Hypotheses

While we are only at the beginning stages in developing a theory on the role of telecommunications in development, we can propose a series of hypotheses concerning the relationship of telecommunications to the development process:

- *The effects of telecommunications use do not accrue exclusively to the users, but accrue also to the society and the economy in general.*

Generally, the benefit of a product or service accrues to the consumer who purchases it. However, the benefits of telecommunications may accrue to others besides the users of the service, because the function of telecommunications systems is to convey information. At its simplest form, the benefit of a telephone call may accrue to both parties, although only one of them has paid for the call, because the purpose of the transaction is to share information. But others who are not even parties to the communication may also benefit from the transfer of information.

In economic terms, telecommunications can be considered as a public good, characterized by the presence of externalities. Thus, it is not sufficient to measure the benefits of telecommunications only by reference to direct user-benefits. The externalities of telecommunications must also be taken into account, i.e., the benefits which accrue to individuals or groups, other than those who pay for the telecommunications transaction.

Lesser and Osberg (1981) cite three types of external economies associated with telecommunications:

1. *Network externalities:* Each time a subscriber is added to the telecommunications network, existing subscribers acquire a greater calling capability. Also, the larger the existing network, the greater the value of joining the system. There is typically a fixed charge for network connection, so that the greater benefit of access to a larger system is not captured by the price system.
2. *Calling externalities:* Each contact via telecommunications benefits the party called as well as the caller. Since the originator typically pays for the contact, the benefit to the party called represents an external economy not accounted for in the transaction price.
3. *Indirect or secondary benefit externalities:* Individuals who do not use telecommunications may benefit if socially or economically beneficial

activities are enhanced through the use of telecommunications. Telecommunications may contribute to increased efficiency, access and/or quality of goods and services.

Thus the benefits of information-sharing using telecommunications may in many cases extend far beyond the parties involved. For example, a call in an emergency from a nurse at a village clinic to a doctor in an urban hospital will benefit the patient who receives treatment as a result of the doctor's advice. Similarly, in an epidemic, the beneficiaries of an emergency call are the inhabitants who receive medicine as a result of the relief efforts coordinated by the call. The use of telecommunications to coordinate transportation of supplies and products may benefit both producers and consumers. The call from a village shopkeeper to order a spare part for a water pump will benefit the farmer who needs irrigation as well as the shopkeeper who sells the part. And a call to arrange transportation of ripe fruit and vegetables to market will benefit the farmer as well as the wholesaler and transporter.

There are two major economic implications of this hypothesis. The first is that there is likely to be a surplus of benefits over costs to the user of a telecommunications system, and these benefits may accrue to both parties in the communication and to others in the society as well. The second related implication is that the revenues derived from telecommunication systems do not adequately reflect the benefits derived from them. Thus, national planners should take into consideration the indirect benefits to the society of telecommunications use as well as projected revenues in calculating whether the returns on investment justify the capital and operating costs of the system.

The following are hypotheses concerning indirect benefits of telecommunications:

- *Telecommunications permits improved cost-benefits of rural social service delivery.*

In countries where a commitment has been made to provide or extend rural services (e.g., health care, education, and agricultural extension services), a key policy issue concerns the cost-effectiveness of alternate ways of providing rural services. Locating highly or moderately trained professional or paraprofessional workers in rural locations may not be feasible because of cost of initial training, the continuing salaries, and the substantial bureaucratic overhead (including communication and transportation) of managing a large, geographically dispersed organization. Even if a sufficient number of highly trained personnel were available and budgets were sufficient to pay their salaries, it still might be very difficult to induce them to live and work in rural locations. If minimally trained rural workers are utilized on either a volunteer or nominally paid basis, then the management and supervision requirements and continuing education requirements can be greatly increased. In the absence of good telecommunications infrastructure, the travel and professional labor costs associated with management, supervision, and continuing education may be prohibitively expensive. But without such supervision and continued training, programs may be significantly less effective. The consequences may be a low quality of rural education, lack of agricultural extension information to subsistence farmers, and poor or nonexistent rural health care. With reliable rural telecommunications, facilities such as telephony or audio conferencing may be used to provide supervision and assistance to rural workers.

Savings in training costs, labor costs, and transportation costs may justify significant allocations for telecommunications services. Dedicated communication systems may be economically out of reach for many social service programs, but such programs may be able to afford shared use of a general purpose telecommunication system supported by revenues from other social service or economic enterprises, as well as by consumer revenues.

- *Telecommunications permits improved cost-benefits for rural economic activities.*

Rural areas that lack telecommunications infrastructure may be at an informations disadvantage (and consequently an economic disadvantage) relative to urban areas. In such cases, urban areas, which usually have more developed telecommunications and transportation substitutes and which are in closer contact with large domestic markets as well as international markets, will have superior access to nonprice economic information. A telecommunications infrastructure may help ease some of the information inequalities which now exist between rural and urban areas.

Timely access to relevant information such as weather reports and prices and availability of necessary inputs (seeds, fertilizers, tools, credit, etc.) should make rural agricultural enterprise more efficient. Timely access to technical agricultural information, such as might be provided by an agricultural extension service, may also improve production. Rural telephony or telex services may improve efficiency of locating and ordering supplies as well as the marketing of the resulting produce. In many developing countries rural population growth has led to a surplus of labor relative to the available agricultural land. Rural urban migration patterns are putting great pressure on urban areas with insufficient jobs and insufficient urban infrastructure (e.g., water and sewage systems). Consequently, development of rural nonfarm enterprise may be necessary. The availability of an underemployed rural labor force at advantageous wage rates may provide incentives for rural enterprise, if reliable telecommunication and transportation infrastructure are available to facilitate the coordination of necessary inputs and marketing activities.

- *Rural telecommunications permits more equitable distribution of economic benefits.*

More important, perhaps, than the direct benefits of rural telecommunications (e.g., in jobs in installation, operation, and maintenance) are the indirect consequences for distribution of economic benefits. Improved cost-effectiveness of social service delivery programs and facilitation of increased or more productive economic activity could contribute to the goal of reduced inequities between urban and rural residents.

In addition, the availability of reliable telecommunications linking rural to urban areas may make it easier for rural people to make their needs known. As a result, they may be more effective in claiming their fair share of national budgets. To the extent that government bureaucracies operate on the principle of "oiling the squeaky wheel," telecommunications infrastructure that permits national governments to better hear the views of rural people may be beneficial to them.

However, the benefits outlined above do not occur in a social or economic vacuum. Therefore, the following hypothesis should be added:

- *A certain level of organizational development and complementary infrastructure is required for socioeconomic benefits of telecommunications to be realized.*

Development may be seen as an awareness-action process in which access to information is critical to enable people to understand problems, evaluate alternatives, plan and act. Traditionally, the mass media have been considered the main purveyors of information, and little attention has been paid to telecommunications.

Cherry (1971) refers to telecommunication services (telephone, telex, telegram) as organizational media (as opposed to the "informational" mass media). We should not expect that people will spontaneously begin to use the telephone to gather information or to form organizational linkages. However, we can hypothesize that the telephone will be used for organizational and informational purposes within the existing institutional framework of the society.

An institutional framework can exist in any type of society, be it a traditional village with chiefs, elders, or counselors; a commune; or an urban collectivity of families with strong neighborhood or cultural ties. We need not be preoccupied with indicators of "modernization" but rather of organization or functional networks.

It can be expected that communication will begin within these networks:

- between members of extended families;

- between tiers of workers in a service organization such as a health system;
- between field staff and administrative staff in development projects.

As scopes of interest broaden and commmunities of interest widen, we can expect the telephone system to reinforce new linkages, for example, between leaders within a region, and between local representatives and central officials. The degree of organizational structure appears to influence a community's or region's ability to use the media for collective rather than individual purposes.

It appears that there are certain societal criteria which must be met before societies are able to use information for collective goals. If people perceive themselves as clusters of families with no links between them except blood ties, they have not reached a point of collective identity where organization is likely to take place, and thus will not use media as tools to gain information needed for organization. However, it appears that a society can reach a "take-off" point where organization becomes a perceived goal, and at that point the media can be used to facilitate this process. Of course, interpersonal and mass communication also play a role in the process by which fragmented or isolated groups come to perceive themselves as part of a larger entity.

Modernization research focuses on defining "modern man's" attributes, finding examples of modern men in underdeveloped settings, and suggesting approaches including use of media which will help to "make men modern." (Inkeles, 1969) However, modernization theory appears to be culturally biased because it presents a western industrialized model of society as a goal and focuses on the individual rather than the society as a whole.

The modern man theory does not seem to be of much help in explaining telephone usage. Rather than look for isolated individuals, we should look for those who play effective roles in the society. Rather than singling out "modern men" who may be sprinkled through most developing societies, it is more fruitful to the understanding of media use to examine the key members of the society—leaders, representatives of institutions, businesses, etc. It is possible to examine their traits, and through them to postulate directions of the society.

The following model is presented as a means of predicting levels of purposive media use in a society, and of determining a "take-off point" at which media use for collective goals begins. It should be considered in conjunction with the section on "public goods theory."

The focus of analysis is on the key leaders/officials of a community or region. The two traits or variables to be examined are "representativeness" and "understanding of the external environment."

By "representativeness" is meant the recognition of the scope of the constituency/institution the person represents. For example, a leader who sees himself as representing only the inhabitants of a village would rank low in representativeness. A businessman with ties to a regional distribution center and a health worker reporting to a regional health center would rank relatively high.

By "understanding of the external environment" is meant sufficient knowledge of the political, economic, and social systems and the common applications of technology of the dominant society or the nation to be able to interact effectively with it.

It is the "understanding of the external environment" concept which parallels some of Inkeles' modernization scales—e.g., "growth of opinion awareness," "openness to new experiences," "particularism–universalism," "efficacy of science and medicine," "efficacy of technical skills." (Inkeles, 1969) However, this definition of "understanding of external environment" implies knowledge and not evaluation of that environment, and thereby differs from Inkeles' notion of modernization. It implies, rather, a sense of pragmatism coupled with the necessary skills to enable the person to acquire from that environment the information/capital/goods/services his people require.

A Type I person (see Table 2.1) represents a small local group of people and knows little of the workings of the external social/economic/political/technical environment.

A Type III person represents a large or scattered constituency or institution. He has little understanding of the "external environment."

A Type IV persons represents a large or scattered constituency or institution and understands the workings of the "external environment" sufficiently to function in it.

It is postulated that this matrix can predict purposive media use by leaders which is likely to be eventually diffused to other members of the society.

A Type I person is likely to use media only for personal reasons.

Type III persons are likely to engage in information-seeking for collective goals. They will be able to use media for these purposes if they are available. However, the scope of this information-seeking will be limited to contact with their peers and relevant individuals on the periphery of the external environment (e.g., local administrators, suppliers, teachers, health personnel).

Type IV persons will engage in information-seeking for collective goals. Their knowledge of the external environment will lead them to search in it (either through reception of mass media messages or through interactive communication) to seek solutions to their problems.

It is hypothesized that category III represents a "take-off point" for purposive media use for collective goals. At this point, the person recognizes that he or she represents a constituency with shared needs and will use communication to seek solutions. However, contacts will be limited generally to his peers and he or she will interact only minimally with the external environment.

Even within an appropriate organizational framework, the benefits of telecommunications use will be limited unless there is at least minimal complementary infrastructure. Most important is transportation.

Of course, there may be instances when communication effectively substitutes entirely for transportation, such as when information in the form of consultation, ordering of supplies, clarifying records, etc. is required. However,

Table 2.1
Variables Influencing Media Use

	Understanding of External Environment	
Representativeness	**Low**	**High**
Low	I	II[a]
High	III	IV

[a] The second category is logically a rare condition. There are likely to be few cases of people who understand the external environment but perceive themselves as representatives of a small unique group. An example might be a "drop-out" who leads a small farming commune. But even they are likely to recognize that their people share attributes with other farmers or communal groups.

in most cases, transportation will be required at some point; for the delivery of the supplies, documents, or medicines, for evacuation of patients, for moving personnel.

Several studies reviewed in Chapter 5 below examine the relationship of telecommunications to transportation. Less obviously, other infrastructure such as power supplies, clean water, and other facilities may be required to ensure maximum benefit from communication systems. Communication can contribute only minimally to improving health care if sanitation and living conditions are unsatisfactory. And communications cannot be expected to support other activities and services unless adequate facilities and equipment are available.

• *Telecommunications use can facilitate social change and improved quality of life.*

Wellenius (1971) proposed that development is related to the amount and variety of interaction. As a society changes from rural to urban, the amount of social interaction increases. He finds that three types of phenomena take place: substitution for alternative means of communication, generation of new communication, and new requirements due to greater interaction.

Telecommunications may help promote social changes by promoting mobility for the population. Cherry (1977) has argued that the telephone acts as a means of mobility—one can move about the country while remaining in the same place through the use of a telephone network. The network offers possibilities for contact with strangers. These network characteristics could tend to make the users more cosmopolitan. This, in turn, leads to a state of "empathy," which Lerner (1958) considers to be the most important element in creating the "modern" personality necessary for development. This argument weakens somewhat in the light of some limited research on the development of new contacts through telephone systems and through face-to-face contacts by Thorngren (1977) which indicated that telephones act more to reinforce contacts which have already been made than to help form new contacts.

Ball (1968) in this regard has argued that as family and friends are scattered geographically by mobility and change, ready access by telephone is made to compensate for the loss of shared environs while facilitating the dispersion. Similar points have been made by Wurtzel and Turner (1977).

An important note in relation to migration patterns in particular and social relations in general, is that the existence of the telephone channel does not cause social changes, but rather allows them to proceed. In this regard Abler (1977), in discussing the dispersal of the American metropolitan areas, noted that

> Rate reductions for long distance calls in the 1950's and 1960's complemented the ongoing evolution of the nation's economy by making it feasible to manage and coordinate widely dispersed facilities. The net effect has not negated the advantages of locating in metropolitan areas, but it has put all these areas on a more equal basis in competing for information activities . . . The telephone network made it possible for dispersal which had other causes, to proceed.

The existence of a telecommunication system may be necessary to support the new types of organizations which occur as part of the development process (see Chasia, 1976). Cherry (1977) has argued that "in the economic sphere, the telephone service is essentially organizational in function; it creates productive traffic."

On a broader level, Parker (1978b) has argued that the availability of two-way channels for communications enhances the growth of knowledge within a nation. Presumably, this would occur because two-way channels may equalize various parties' access to sources of information.

However, as noted in McAnany et al. (1980), the usefulness of access to information in rural areas must take into consideration the structural constraints on the uses of information which may come through a telecommunications system—access to information provides no assurance that the information can be used because of the existence of power hierarchies and limited social organization. It is probable that many of the problems of rural development are problems of distribution of resources and power. As such, telecommunications development may have little impact on such problems, and indeed, could exacerbate them under certain conditions by increasing the resources available to urban and rural elites to exploit the rural poor.

Parker (1976) has argued that telecommunications may both reflect and reinforce the social structure of a nation. Parker views institutional structures as being composed of communication patterns which may be shaped by the communications technology available to society:

> The communication technology of a society determines who can speak to whom, over what distances, with what time delays, and with what possibilities for feedback or return communication. This is the heart of what is meant by social organization. It makes less sense to say that the social organization is caused by the

pattern of communication interactions in the society than it does to define the social structure in terms of the patterns of communication (including order-giving). The culture of a society can be defined by the messages that are transmitted in these social patterns. The messages of a society are obviously shaped by the media they are transmitted through, as well as being creations of the institutional structure of society. Therefore, careful attention should be paid to the form of the communication technology installed in support of development.

The existence of telecommunications services may facilitate political development. Teer (1975) has argued that telecommunication spreads participation within a nation. Myrdal (1970) has suggested that most developing countries profess egalitarian ideals. However, such ideals may need pressure "from below" to be realized—telecommunications may provide a channel for such pressure.

Simpson (1978) points out that successful efforts to provide communications services to remote regions may result in greater participation by their residents in the society. He states concerning Canada:

> The predominantly south to north communication links of the past are becoming truly two-way communication links. The voice of the northern society is being heard in the south ever more frequently and with greater clarity.

Goldschmidt (1978) reports on the impact of communications on Atka, Alaska, a community of approximately 88 people located on an island in the Aleutian Chain. This native (Aleut) village subsists generally from fishing, and more recently, from the limited employment generated by a coast guard docking facility. The community has no air-strip. Prior to its receiving a satellite earth station, all communications with the outside took place via mail which was exchanged once a month when a mail barge visited Atka from Adak. Such slow communication made application for government grants virtually impossible.

With the satellite earth station, daily transactions with the Anchorage office of the Aleut League, the organization representing Atka in most of its government dealings, have helped generate increased government grants-in-aid programs to Atka. The turnaround time for grant applications, including the simple act of notifying the village that grants are available, has been reduced from a minimum of one month to just a few minutes. Also, the possibility of instantaneous communications has permitted more timely ordering of supplies, educational materials, and health goods which are shipped in on the mail barge.

It is important to note here, however, that the increase in economic and political activities in villages like Atka in Alaska was not caused by the introduction of telephone service. Rather, the catalyst was in the Alaska Native Land Claims Settlement Act of 1971. The Act precipitated a series of changes in native life as it resulted in extensive organization of the Alaskan natives and the transfer of large amounts of money and land from the U.S. government to the natives through various native organizations. The telephone link expedited these changes and probably allowed certain changes which might not otherwise have occurred given the poor communications prior to satellite service. However, the telephone as a facilitator of social change must be distinguished from the telephone as the cause of social change.

Hudson (1974, 1977) reports that the communication system is considered important for regional development by the Indian people of northern Ontario who cited communication as their top priority for many years. In a region where travel is difficult and expensive, the telephone allows leaders to plan, discuss priorities, and coordinate strategies. Previously, the village chiefs had no way of coordinating their planning until they arrived at meetings with government or commercial agencies—which clearly placed the agencies at a strategic advantage.

The telephone system can be a vital tool to enable people to participate in their own development. However, as already noted above, it appears that certain preconditions must exist: there must be some level of local organizational structure, and there must be some perceived community of interest with other parts of the region or subregion. For example, chiefs who represent their communities and share common concern about the development of their region will use communication as an organizing tool regardless of their education or their ability to speak the national language. But, unstructured village groups who have little awareness of common concerns with other communities are likely to use the telephone purely for social communication between family members and friends.

It appears that introducing telephone service can improve the quality of life in remote villages. For example, existence of private line and toll circuits to Alaskan villages has allowed the continuation of regular consultations between village health aides and medical centers initiated during the ATS–1 project. It has also allowed easier contact between village residents and medical services. For example, in Alaska, the amount of time women from the Aleutian islands must spend in Anchorage receiving pre- and post-natal care has been reduced from two months to one month—the women can remain with their families longer and consult with the Anchorage Native Medical Center by telephone (Goldschmidt, 1978).

The existence of toll service allows continued contacts among native families which have been dispersed due to exigencies of boarding schools, employment, health, and the like. More importantly, this extended contact with families and friends acts to provide a constant flow of information to the villages.

The introduction of exchange telephone service has also had beneficial effects on village life in Alaska and northern Canada. During the winter, telephones have generally made it easier to determine where children are, and whether hunting parties have returned. It also allows the transaction of local business and services without having to go outside to made a personal visit. This is no small advantage even if the village is relatively small.

Native organizations from northern Canada have argued that communication in the isolated northern villages is a life and death matter in interventions before the Canadian Radio-Television and Telecommunications Commission (CRTC) about the quality of telephone service in their region. In 1977 the CRTC sat for 35 hours in four northern communities where they heard some 54 intervenors representing over half the Inuit (Eskimo) population complaining about the telephone service and explaining the importance of communications for emergencies, delivery of medical and legal services, and maintaining contact between dispersed families and friends.

Billing data are an indicator of the importance of phone service to the Inuit and Indians in the remote north. In both northern Canada and Alaska they spend more than three times as much as their urban counterparts on toll telephone service (Hudson, 1981). These data indicate that, from the vantage point of rural consumption expenditures, telecommunication plays a relatively important role in promoting the quality of rural life.

Part II

Findings From the Research

Chapter 3

Telecommunications and National Development: Macro Level Studies

3.1   Introduction

The following chapters present a review of the literature and a synopsis of recent major studies on the role of telecommunications in development. Much of this work was supported by the International Telecommunications Union and the Development Centre of the Organization for Economic Cooperation and Development (OECD). These studies shed considerable light on the hypotheses outlined above. However, they leave many theoretical issues still to be addressed, and many steps still to be taken to move from case studies to generalizable research findings.

The literature review conducted by Hudson et al. (1979) revealed that there had been more written on the topic of telecommunications and development than might have been expected. Writing, however, did not imply research. Many of the articles drew attention to questions of the relationship between telecommunications investment and development, rather than proposing answers. Others critiqued techniques used in the past, some suggesting different techniques, but not applying them. The research that did exist tended to fall into two categories:

- analysis of historical data at the national level to identify a close relationship between economic development, as measured by key national development indicators, and investment in telecommunications;
- case studies of the applications of telecommunications in various sectors including health services, education, agricultural production and marketing, fisheries, primary industries, etc.

The backgrounds of the authors tended to cluster in a few major categories:

- engineers, often from telecommunications administrations or planning agencies concerned with telecommunications, and a few from universities;
- economists from the telecommunications industry, the development banks, and from universities;
- social scientists other than economists, with backgrounds in political science, anthropology, social psychology, etc.

Their articles and studies were found in:

- social science journals;
- technical journals (even for articles on socioeconomic impacts of the technology);
- conference proceedings, generally of technically-oriented societies or agencies;
- unpublished research reports from technical organizations, development agencies, and government departments.

Thus, many of the documents collected for the review were not commonly available.

The authors and researchers did not seem to constitute an "invisible college." With few exceptions, their interactions seemed to be within their own primary discipline—engineering, economics, sociology, etc. They also appeared geographically and culturally isolated. Western writers were not familiar with research conducted in the U.S.S.R. and Poland, for example; American researchers had not been exposed to studies conducted in France. Writers concerned with developing countries were unfamiliar with a substantial body of literature on telecommunications services in remote parts of North America. And researchers concerned with the impacts of telecommunications on energy consumption and productivity seemed to limit their research and predictions to the industrialized world, without considering what implications they might hold for the developing world. Yet in the last 5 years a network of technical and social scientists, planners, and policy analysts has begun to form. Growing institutional interest on the part of the ITU, the OECD, bilateral aid agencies, and the development banks in the indirect or developmental benefits of communications has been one factor. The advent of satellites, which through demonstrations, experiments, and operational services have proved their utility in reaching previously unserved rural areas, has brought technical and applications planners together. And major policy developments such as the World Administrative Radio Conference of 1979 and calls by developing nations for a new world information order and new economic order have highlighted the role of communications in the development process.

The following chapters will present the findings of selected studies and brief references to other studies which have contributed to our knowledge of the role of telecommunications in development. The purpose of this review is to bring to the attention of the reader works from a wide variety of sources that may shed light on the role of telecommunications in the development process either through the results of empirical research or through theoretical insights. The review is illustrative rather than exhaustive, as research in this field has increased significantly in recent years. However, the following chapters and the bibliography should provide a guide to the range of theoretical and empirical research in this field.

3.2   Problems with Earlier Studies

The GAS-5 Committee of the CCITT has published a series of reports on the effects of telecommunications on national economies (see CCITT, GAS-5 "Economic Studies at the National Level in the Field of Telecommunications 1964–1976"), which have become increasingly sophisticated in their treatments of the probable effects of telecommunications on national economic development. While rural development is treated only marginally in these reports, it is still useful to note the various functions imputed to telecommunications which may prove suggestive in future studies.

GAS-5 views telecommunications primarily as a means of diminishing the costs of time and travel:

> Telecommunication is, in a sense, a substitute for presence. Its availability allows the user to overcome some of the disadvantages of distance. As the next best thing to being there, telecommunications can substitute for travel with an associated saving in time, cost and personnel. As a substitute for the mails, telecommunications can act to save time and to speed up a decision-making process. As a means of access to information, telecommunications can allow quick and frequent access, by managers and other users in a remote area, to the pools of information and talented advisers found in large population centers. In the future there are even greater prospects of substituting traditional forms of infrastructure with specific types of telecommunication infrastructure. (1972, XII: 5)

These benefits are manifested in several ways. Among the functions of telecommunications reviewed by the ITU in the GAS-5 studies are:

1. Importance for secondary manufacturing and the tertiary (government, finance and services) sector;
2. Substitution for travel; potential energy savings;
3. Decentralization of business and industry through capability to transfer information quickly and accurately;

4. Benefits to consumers in providing information and facilitating accurate ordering and delivery of goods;
5. Maintenance and expansion of tourism which in turn expands the service sector;
6. Increased efficiency and geographic coverage for government administration and delivery of services;
7. Organizational impacts on agricultural production through improvements in ordering and delivery of supplies and equipment, more timely access to services and increased availability of marketing information.

The GAS-5 studies suffer two major deficiencies. First, most of the discussion is nonempirical. While the hypotheses about what telecommunications may do for an economy seem reasonable, there are insufficient data at present to validate any of them.

Second and more important for this book, the research and writing are almost exclusively oriented towards the urban, industrialized nations. The GAS-5 has downplayed the importance of telecommunications for agriculturally-oriented nations with lower GDPs—telecommunications is presented as an essential component of mass consumption societies. Countries which are largely agrarian with low GDPs depend largely on primary sector industries which, due to the simplicity of the production process, require little telecommunications. Also, the lack of a well-developed urban economic sector ostensibly implies that there is little need for the types of telecommunications seen in more industrialized nations. While this argument may reflect the actual distribution of telecommunications among nations, it is important to note that the technology for telecommunications has primarily been oriented towards urban usage and, until recently, telecommunications was assumed to be more important for urban than rural areas. It is not certain given these circumstances whether the low level of rural telecommunications reflects differences in the needs for certain types of electronic communications between the urban and rural sectors, whether it reflects the lack of a rural-based technology until recently, or whether a series of self-fulfilling prophecies relating to long-standing political divisions between urban and rural areas created a situation where rural areas were denied facilities, or were provided with facilities of limited accessibility due to poor quality or high prices.

The reference made to rural development by GAS-5 is directed primarily at fairly developed agricultural sectors and certainly has little to offer peasant-based agricultural sectors. In large part, this results from lack of any experience with rural telecommunications in the LDCs. However, the GAS-5 reports consistently indicate a bias towards the capitalist industrialized economies.

In 1972, Wellenius stated that most efforts at connecting telecommunications to national development are done at the aggregate level, through elaboration of independently derived data. He found no systematic approach to cost benefit analysis (Wellenius, 1972). Such is still the case.

The major problem in past research on the statistical relation of telecommunications and development concerns the inferences which may be drawn from the analyses. Most studies which have explored the role of telecommunications in development suffer from being correlational in nature. White points out that correlation does not equal causality. He cites the example that there could be a correlation between the incidence of crime in certain areas on London and the density of Salvation Army workers (White, 1971). In that case underlying variables such as poverty, unemployment, and/or poor living conditions may explain the presence of both crime and Salvation Army workers. In terms of development, high GDP may in fact lead to telecommunication investment, and we have learned nothing about whether telecommunication could have contributed to economic development if it had been available.

A typical procedure is to correlate telephones per 100 population with GNP or GDP per capita. This correlation is normally quite high. Studies done by Marsh (1976), Shapiro (1976), and the CCITT (1968, 1972), among others, incorporated such a procedure in their analyses. From that point, the factors relating to demand for telecommunications services within a nation are then examined. For example, Shapiro states: "A nation's industrial development is related to the scope and quality of its telecommunication system." He correlates telecommunication density to GDP, finding an "r" of .70 for Latin America (Shapiro, 1976).

Unfortunately, the various indicators of telephone investment, density, and the like are not generally useful for linking telephone development with rural development as these figures fail to indicate the access of the population to the service, the distribution of the service and, most importantly, the reliability of service. Statistical indicators often are heterogeneous, as with the case of telephone density where the figures for the industrial nations reflect demand, while for the LDCs they reflect supply. The point about reliability is particularly important as, without some notion of reliability (e.g., completion ratios), we cannot determine if the investment in telecommunications stimulates the economy as any capital investment would, or whether there is some economic property derived from the communications services provided by the investment. Given the generally acknowledged unreliable quality of telephone service in most of the LDC's, we may speculate that the former case is more important; it is possible that telephone service has only promoted economic development in nations which can economically sustain the high cost of maintaining reliable service.

Also, no causal inferences may be made from these studies even assuming that telephone service is reliable. One does not know the degree to which telecommunication is influencing development or development is influencing telecommunications. Such a correlational approach does not allow one to specify the direction or magnitude of telecommunications' contribution. Looking at factors influencing telephone demand may be constructive for planners in establishing forecasts and investment programs. However, it does not touch the question of what roles telecommunication plays in rural development.

Two additional studies take the correlational approach a few steps further. Gillings (1975) performed a cross-sectional analysis of twenty-nine nations to test a hypothesized catalytic effect of telecommunications upon economic development. Formally stated, his hypothesis was that per capita GDP of secondary and tertiary sectors of a national economy is a function of the catalytic effect of telephone use and availability acting upon factors supporting the development of these sectors. This catalytic effect is hypothesized because telecommunication increases the flow of information within society. Increased information flow can permit more rapid decision-making. It also results in greater efficiency of industrial operations. Thus, these factors should combine to create greater economic output. This reasoning leads to the corollary that the catalytic effect is higher in nations of higher information flows.

Gilling's results suggested that increases in developmental support factors will be less effective in improving economic performance without a sufficient level of telecommunications use and availability. Also, additions of telephones without certain necessary minimum levels of capital formation and quality of manpower will be less effective than if telephones increase appropriate to the level of support factors.

While Gilling's hypothesis seems to imply that telecommunications mediate the influence of support factors on development, his results do not support his conclusions. In a three variable system, a number of different models can be made to relate the three variables. A reduction in correlation between two of the variables when controlling for a third would support models other than the one which Gilling proposes. For example, a model with development causing telecommunications and support factors, and telecommunications thus causing support factors would also be supported by Gilling's partialling exercise. Another problem with the analysis is that it fails to take into account the actual distribution of the economic benefits, or distinguish between urban and rural components in the countries studied. It must be noted however, that data on urban and rural telecommunication systems, rather than national data are notoriously difficult to get.

Cruise O'Brien, et al (1977) have completed a rather extensive description of telecommunication in relation to development. GNP per capita, as well as a number of communication variables and social indicators were used throughout the study. A composite index of the communication variables was created through principal components analysis. The communication indicators were newspaper circulation per 1000 population, newsprint consumption per capita, telephones per 100,000 population, radios per 1000 population, and television per 1000 population. Data for 73 nations from 1960 and 1970 were used.

Several major findings are of interest. Changes in the amounts of communication variables from smaller GNP deciles to larger ones were examined. Telephones expanded markedly between $440 and $730 per capita and again above $1,400 per capita. However, telephones did not expand as rapidly over the whole GNP range as most of the other communication indicators.

Growth in GNP per capita, and the communication variables between 1960 and 1970 were then examined. Five groupings of nations based upon GNP growth were made. It was found that telephones showed a steady growth rate from the lowest groups. In countries with high growth rates over this period telephones appeared to be important, as well.

Changes from 1960 to 1970 in GNP and the different media for nations at different levels of GNP showed that telephones have increasing growth rates at higher levels of GNP. Nations with high levels of GNP and higher levels of GNP growth, then, show higher growth in telephones.

Looking at the standardized values of communication variables at different levels of GNP for 1970, the study found that telephones in nations at lower GNP levels are at higher levels relative to the other media. In nations at higher GNP levels, the levels of telephones are lower relative to other media. Thus it would seem that nations at lower GNP levels have concentrated more on increasing media such as radios and newspapers. Nations at higher levels of GNP have concentrated upon increasing the numbers of telephones.

It is difficult to ascertain what underlies these relationships. It is possible that a higher level of GNP is required to make full use of the economic potentialities of the telephone. It may also be the case that if nations at lower levels of GNP were to concentrate greater investments into telephones, GNP growth rates might increase. Unfortunately, we again have the problem of inferring whether these high rates of growth in telecommunications were being caused by growth in development, or vice-versa.

Several other studies have examined the relationship between GNP or GDP/capita and telephone density. Regression equations have been developed by Phuc and Dennery, Jannes and Montmaneix. Using data from the CCITT for 30 countries, Montmaneix (1974) found that long distance traffic was growing more than twice as fast as GNP. Dormois and Gensollen (1976) use a different technique to quantify telephone development with surveys about household phones and individual usage patterns. Variables include income, socioeconomic status, and equipment usage charges. Their studies do not include business phones.

Through analysis of input/output tables, Bernard (1976) suggests that for developed countries telecommunications represents final consumption, or an effect of wealth, rather than a cause, or factor of production. Bernard's technique involves construction of a hierarchial taxonomy of activities using the criterion of the degree of cause or effect of wealth on each activity.

3.3   The Question of Causality: New Evidence

As noted above, many studies on the relationship of telecommunications to development have used a correlational approach, but have failed to determine the direction of causality once a positive correlation between investment in telecommunications (or some proxy measure) and economic growth has been established.

Using a more sophisticated approach to correlational analysis than previous researchers in this field, Hardy (1980, 1981) developed hypotheses designed to answer several fundamental recurring questions:

1. Does telecommunications contribute to economic development, or does economic development contribute to telecommunications, or both?
2. Do business telephones contribute to economic development in developing countries? In industrialized countries?
3. Do residential telephones contribute to economic development in developing countries? In industrialized countries?
4. Does the telephone contribute more to economic development than a mass medium such as broadcast radio?
5. Does telecommunications contribute more to economic development in the industrialized countries than in the developing countries?

The indicator Hardy used for economic development was Gross Domestic Product (GDP) per capita. A second development indicator, energy consumption per capita, was also used in some analyses. Telecommunications was measured in telephone density (telephones per hundred population). Data were analyzed for 37 developing countries and 15 industrialized countries over a 14-year period (1960–1973).

Hardy used cross-sectional, time-series regression analysis, a technique that enabled him to determine whether and to what extent one variable would lead or lag another by comparing data for each variable for one year (time t) and for the year previous (time $t-1$). The use of time series data and cross-lagged correlation techniques allowed Hardy to identify causal relationships in the statistical sense. The results of this analysis shed considerable light on questions of causality in telecommunications and economic development.

There is a positive correlation between telecommunications (measured in telephones per 100 population) and two economic indicators (per capita GDP and per capita energy consumption). In the industrialized nations, the contribution of telecommunications to economic development appears to be somewhat weaker than in the developing countries.

The correlational relationship does not seem to exist in the case of radios; the number of radios per 1,000 population appears to be unrelated both to number of telephones and to economic development. It thus appears that two-way telecommunications facilities such as the telephone have a greater influence on economic development than a one-way medium such as radio broadcasting.

Hardy produces some interesting results concerning the comparative impact of business and residential telephones. Contrary to what might be expected, investment in business telephones does not appear to cause or lead economic development in developing countries, while this relationship is found in the industrialized countries. This may be because the penetration of business telephones in many countries is very limited, or this category may not be so commonly used because of differential tariffs in installation and rental charges for business and residential telephones. In developing countries, according to Hardy, it is residential telephone penetration which leads development. This finding suggests that, contrary to the conventional wisdom based on experience in industrialized countries, the residential telephone may be an important component of developmental infrastructure in developing countries. This finding can be explained by a number of factors, one of which is probably the fact that a very large proportion of so-called residential phones are used for business communications, by, for example, merchants, professionals, and cottage industries which generate employment and contribute to economic growth while requiring only limited capital investment.

Hardy (1980, 1981) used the methodology he had developed for his analysis of the relationship between economic development and telecommunications investment to attempt to determine the measurable effect of telephones on economic development.

Hardy used two approaches. The first approach involved a regression analysis of the variables used by Denison (1974) in his investigation of the role of knowledge in economic growth. The second approach is based on the coefficients Hardy developed in his own study. The first approach suggested that a 1% rise in the number of telephones per 100 population between 1950 and 1955 contributed to a rise in per capita GDP between 1955 and 1962 of approximately 3%.

The second approach to analysis measured the multiplier effects of the telephone on economic growth at different stages in the development of the telecommunications system (measured in terms of telephone density) and measured the percentage gains in per capita GDP resulting from a given increase in telephone density. The data resulting from these two methods are summarized in Figures 3.1 and 3.2.

These figures show that the contribution of the telephone to economic development is particularly high in countries with very low telephone density (less than 10 telephones per 100 population), which tend to be the poorest countries. Thus, the lower the telephone density, the greater the potential contribution of telecommunications investment to economic development, measured in increase in GDP per telephone. Thus rural and remote regions as well as the poorest nations may stand to benefit from telecommunications investment. However, see the caveats at the end of this section.

The question of who benefits from investment in telecommunications has received little attention. In further research, Hardy (1980, 1981) tested the hypothesis that the structural properties of interactive communication facilitated by telecommunications will contribute to more equitable income distribution in a

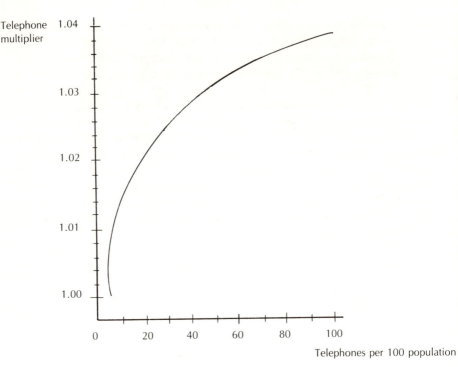

Figure 3.1
Telephone Multiplier as a Function of Telephones Per 100 Population
(Hardy, 1981)

nation. Hardy used World Bank data showing percentage of national income received by the wealthiest 5% of the population, the poorest 20%, and the remaining middle 75%, during the decade from 1960 to 1970.

The results indicate that increase in telephone density is related to the redistribution of wealth to the middle income groups of a nation from the highest income groups, while the lowest income groups are unaffected by telecommunications growth. Thus, telecommunications appears to contribute to a shift in income distribution away from the wealthiest group in a society, and perhaps fosters the development of a middle class.

These findings from the Hardy study can be summarized briefly:

1. The telephone appears to be a much more important factor in the development process than one-way communication systems such as radio.
2. The role of the telephone in economic development appears to be more important in the developing countries than in the industrialized countries.
3. In the developing countries, the residential telephone is far more economically important than it was generally thought to be.

4. The lower a country's level of development, the greater the potential contribution of telecommunications to economic development.
5. Growth in telecommunications appears to benefit primarily the middle income segment of the population.

Several qualifications should be made relating to the Hardy study. The analysis used available data and indicators which may not be the best for our purposes. They are national indicators and tell us nothing of distribution within a country. Income is not likely to be evenly distributed throughout a society. The number of beneficiaries of economic growth may be very small. Telephones are likely to be clustered in cities, so that rural telephone densities are many times lower. Other indicators, such as growth in investment in telecommunications or growth in message traffic, might be more useful. The analyses focus on short term year-to-year relationships, and may not reflect longer term effects.

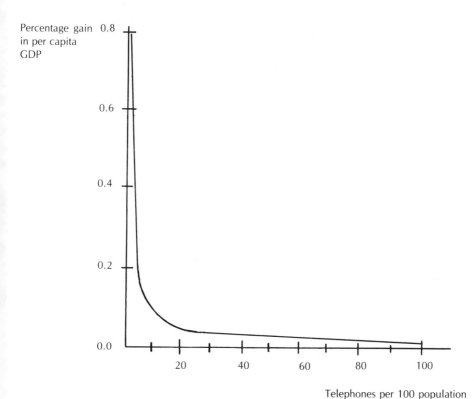

Figure 3.2
Percentage Gain of Per Capita Gross Domestic Product as a Function of Telephones Per 100 Population (Hardy, 1981)

Although the results indicate a greater contribution of telecommunications to economic growth for poorer nations, there may be some threshold or take-off point below which the country or region simply does not have the resources in terms of capital, trained workers, and infrastructure to apply telecommunications constructively. For example, installing telephones in a semi-arid region sparsely populated by nomads living at a subsistence level would likely contribute little to economic development of the region. And, although the results of the Hardy study give us some important insights into the role of telecommunications in national economic development, they cannot be applied to individual countries or regions.

The question of the comparative contribution to development of mass media and interactive services such as the telephone also requires further investigation. Hardy's research does not imply that mass media do not contribute to development, but the relationship may be more complex than for telecommunications. Factors such as content, message design, target audience, follow-up, etc., all of which are familiar to media researchers, are likely to influence the impact of the mass media. Carefully designed and executed media projects have demonstrated developmental impact. However, these projects constitute a very small percentage of mass media activity. Perhaps the importance of these and other intervening variables explains the lack of a causal statistical relationship between mass media investment and economic development. In addition, the role of the mass media may more often emphasize social and cultural development rather than specifically economic development. Thus other indicators than GNP are needed to measure the developmental effects of mass media.

3.4   The Impact of Telephones and Satellite Earth Stations on Rural Development

Two studies in the ITU/OECD series have applied Hardy's (1980, 1981) methodology to address questions important for telecommunications planning and the allocation of national development budgets. Hudson, et al. (1981, 1982) attempt to determine the economic impact of telecommunications investment in rural regions within a country. They also derive a dollar estimate of the impact of the economy per telephone or per satellite earth station over a multiyear period. In a separate study, Parker (1981) applied the Hardy methodology to estimate the impact on U.S. economic growth of the investments in rural telephone systems by the U.S. Rural Electrification Administration (REA).

In addition to estimating the impact of telecommunications on national economic development, it would be useful to estimate the impact of installing telecommunications facilities in rural regions of developing countries. Communication satellites optimally designed for thin route service may be the least cost means of providing basic telecommunications to rural and remote regions of many developing countries. It would thus be very useful to be able to estimate the economic impact of installing small earth stations for telephone service in rural areas.

Also, developing countries are diverse in their levels of existing and projected investment in telecommunication. Therefore, it would be appropriate to develop a technique to estimate impact on economic growth for regions with different telephone densities and for different rates or telecommunications growth.

The study by Hudson et al. (1981, 1982) undertook these three tasks using the methodology developed by Hardy. Since at present data on telecommunications investments and economic output (GDP) are generally available only at the national level, two approaches were used to estimate economic impact of tele-communications growth on regions with very low telephone densities such as would be found in rural and remote areas of developing countries. One approach extrapolated from results of applying the model to existing data. A second approach specified parameters of a hypothetical rural region and applied the model to that case.

To estimate the impact of telecommunications in rural or isolated areas within LDCs, 30 developing countries in the data set were broken into three groups of ten nations each according to telephone densities. However, even the lowest group had much greater telephone densities than would likely be found in rural areas of many developing countries. Therefore, a hypothetical fourth group was created to more nearly approximate these conditions, with the following parameters:

- Telephone Density: .01 (1 telephone per 10,000 population)
- Population: 10,000,000
- GDP per capita: $100

The model estimates the impact on national GDP of installation of additional telephones. An average growth rate of telephone installations of 8% per year derived from the data was used to determine the number of telephones to be added each year. To provide comparative data on the aggregate and per tele-phone impact of accelerated telephone installation, rates of five times and ten times average growth of telephones were also used.

It is assumed that in many rural and remote areas without a telecommunica-tions infrastructure, the most cost-effective means of providing basic telecom-munications will be by satellite, using small earth stations which would operate with a satellite optimized for thin-route domestic service. It is assumed that an average of 10 telephones would be connected to each earth station. These might be public call offices and telephones located in government offices, clinics, stores, local industries or businesses, etc. in one community, or public tele-phones in villages surrounding the central community and linked to its earth station via VHF or UHF radio.

The design life of the satellite is specified as 10 years. Thus the data are organized to show:

- Impact on GDP of all telephones installed over a 10-year period;
- Average impact on GDP per telephone or earth station (servicing 10 telephones).

These data are purely illustrative. The planner could choose to vary:

- The number of telephones installed;
- The period over which to estimate impact on GDP;
- The number of telephones linked to each rural earth station.

The results of the analysis show the impact of installation of thin route satellite earth stations for three groups of developing nations and for a hypothetical rural region. Table 3.1 shows the average contribution to the national GDP per earth station installed over the 10-year life of a satellite. A range of low, medium, and high estimates is given for each national group to indicate the use of the upper 95% confidence value, the simple regression coefficient, or the lower 95% confidence value in estimating telephone or earth station impact on GDP. The medium range estimate is to be used for purposes of comparison (Table 3.1).

The impact on GDP per earth station increases dramatically as telephone density decreases. Thus for regions of telephone density of 1 telephone per 100 population (represented by Group 2), each earth station contributes an average $22,000 to the national GDP over the 10-year life of a satellite. However, where

Table 3.1
Impact on National GDP Per Thin Route Earth Station[a]
Installed Over Ten Year Satellite Life (Average Growth Rate)

Nations	Impact on National GDP per Earth Station		
	Low Estimate	**Medium Estimate**	**High Estimate**
Group 1 (telephone density 7.6)	$ 4,000	$ 10,000	$ 17,000
Group 2 (telephone density 1.0)	10,000	22,000	39,000
Group 3 (telephone density .2)	30,000	66,000	108,000
Group 4 (telephone density .01)	185,000	367,000	550,000

[a]For this and all subsequent tables, one thin route earth station is assumed to provide service for 10 telephones.

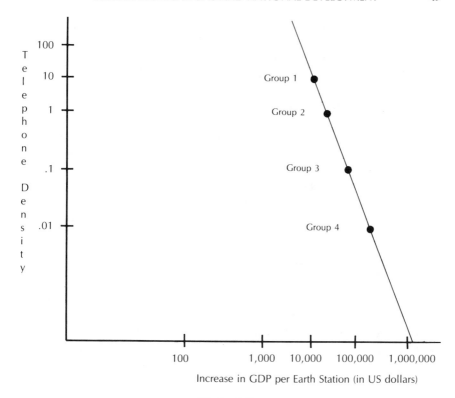

Figure 3.3
Impact on GDP Per Thin Route Satellite Earth Station Installed Over Nine Year Satellite Life (Average Growth Rate and Medium Estimate)

telephone density is as low as .01 or 1 telephone per 10,000 population, each earth station contributes about $367,000 to the national GDP.

Figure 3.1 shows these results in the form of a graph. The relationship between telephone density is logarithmic, and the hypothetical case falls on the line produced by extrapolating from the three groups. Using this graph, the telecommunication planner can estimate the impact on national GDP per earth station for regions of varying telephone density. The planner can simply locate the density of the region on the vertical axis and read off the impact on GDP from the horizontal axis. To find average impact per telephone, the value on the horizontal axis is divided by 10. If the planner wishes to assume that each earth station serves 100 telephones rather than 10, the value is multiplied by 10.

The study also shows the impact per earth station if installation of telephones or earth stations is accelerated to five times the current average growth rate and ten times that rate. The impact per earth station declines slightly as the growth rate is accelerated.

The model estimates the impact on national GDP of telecommunications installations in regions with very low telephone densities. The graph in Figure 4.1 can be used to estimate these economic benefits for average telephone growth rates. A planner can also vary assumptions of growth, telephone density, and initial GDP. The specifics can then be compiled as a special case for analysis using the model as was done for the hypothetical rural region shown in Table 3.2.

The model presented in this study can be used to estimate the impact on national GDP of installations of telephones and/or thin route satellite earth stations in regions of very low telephone density for which no separate telephone density or economic data exist. It should therefore be a useful tool for telecommunication planners who wish to estimate the economic benefit of installing thin route telecommunications in rural and remote regions of developing countries.

3.5   Benefits of Rural Telecommunications in the United States

Rural telephone services have been subsidized in many industrialized countries including the United States, Canada, Sweden, and Finland. However, government subsidies tend to be based on assumed benefits of providing telecommunications services in rural and remote areas. Often these benefits are perceived to be primarily social rather than economic, i.e., improving the quality of life of rural residents. However, the Hardy (1980, 1981) study indicates that there may be major economic benefits in terms of contribution to natural eco-

Table 3.2
Example of Benefits Accrued Over Ten Years for Hypothetical Rural Region[a]

Year	GDP	Increase in GDP Due to Telephones	Number of Telephones
Initial	$1,000,000,000	0	1,000
1	103,711,000	$ 515,000	1,080
2	107,586,000	1,031,000	1,160
3	111,627,000	1,579,000	1,240
4	115,839,000	2,075,000	1,320
5	120,228,000	2,609,000	1,400
6	124,799,000	3,115,000	1,480
7	129,557,000	3,715,000	1,560
8	134,509,000	4,290,000	1,640
9	139,659,000	4,882,000	1,720
10	145,015,000	5,493,000	1,800
	Total (rounded)	$29,300,000	

[a]Increase in GDP per telephone = $37,000; or per earth station = $370,000

nomic growth derived from investments in rural telecommunications. Parker (1981) applied the methodology developed by Hardy to analyze the impact of government loans for rural telecommunications on economic growth in the United States.

The Rural Electrification Administration (REA) was established in the United States Department of Agriculture (USDA) during the Depression to facilitate the electrification of rural America through low-cost loans. Funding was also advanced to install and upgrade rural telephone service. Over the last 40 years, the REA has advanced approximately $3.9 billion in funds at reduced interest rates through its Telephone Loan Programs. These loans have been used to provide initial telephone service for almost 2.9 million subscribers in the rural areas of the United States and to improve telephone services for over 1.6 million subscribers. REA borrowers operated some 6.3 million telephones, or 3.6 percent of the total number of telephones in the country in 1979.

Parker (1981) estimates that without the REA loans, the telephone density in the U.S. would be reduced by 1.1 telephones per hundred population (based on assumptions that subscribers would not otherwise receive service, and that each subscriber accounts for only one telephone).

Hardy's methodology suggests that an increase of one telephone per 100 people in countries with a telephone density of 80 sets per 100 people contributed to an increase of .01% in GDP the following year. The same increase in areas with a telephone density of only 20 sets per 100 people led to a much higher increase in GDP of .09%. Using these estimates, Parker calculates that the reduction in the total stock of U.S. telephones resulting from the absence of the REA Telephone Loan Programs would have caused a loss of approximately .011% of U.S. GDP (given the current telephone density in the U.S. of approximately 80 set per 100 population). In other words, if the 1979 stock of telephones in the U.S. had been lower by 1.1 set per 100 population, U.S. GDP in 1980 would have been reduced by about $283 million (of a total of $2576.5 billion). This "loss" in GDP is, of course, not static; it would increase each year. However, Parker does not project these estimates, as the methodology developed by Hardy focuses only on changes in GDP occurring in the following year.

This potential loss of $283 million in GDP can represent the economic benefit of the REA Telephone Loan Programs. This figure almost certainly underestimates the total benefits for several reasons. First, the effective number of telephones per subscriber served by the telephone companies which receive REA loans is 1.56, and not 1.00 as assumed in this calculation. Second, the telephone density in the rural areas served by the REA is lower than the nationwide ratio of 80 sets per 100 population. Also, only the benefits occurring in the year following the investment were taken into account. And finally, this calculation does not take into account the loans for improving existing telephone services (as opposed to installing telephones for the first time).

To estimate of the upward correction needed to take into account the full benefits of the REA Telephone Loan Programs, Parker calculates that the $283 million estimate would have to be increased to approximately $850 million in 1980, assuming that the telephone density in the areas served by the REA is 60 sets per 100 population, rather than the national average of 80 sets.

Benefits must be weighed against costs to get an idea of the efficiency of investment. The total cost of the REA Telephone Loan Programs to the U.S. Government is difficult to measure precisely. Parker estimates a total cost of $130 million in 1979, based on REA loan data. Parker concludes by comparing this figure with the $850 million in expanded GNP resulting from the REA programs in the following year, that the government's investment has been exceptionally profitable: the benefits to the national economy are 6 to 7 times higher than the cost to the government.

Yet even this ratio underestimates the total benefits of the REA loans. The REA telephone borrowers paid $196 million in taxes to local, state, and federal government authorities, i.e., 50% more in tax revenues alone than these programs cost the federal government. Furthermore, the loans indirectly helped to stimulate economic activities in the rural areas, and hence to bring in additional tax revenues.

Parker (1981) draws three major conclusions from his analysis. First, there are demonstrable benefits for regional and national economies from investments in rural telecommunications facilities, in terms of increased efficiency in the organization of economic activity leading to increased productivity. He estimates the benefits of these programs to the U.S. economy at 6 to 7 times their costs. Second, there are demonstrable benefits for the quality of life in regions and nations from investments in rural telecommunications facilities, in terms of increased availability and cost-effectiveness in the delivery of social services such as health and education. Finally, because these benefits are to a considerable extent external or public rather than private, government intervention is required to ensure optimum investment in rural telecommunications facilities.

3.6   The Economic Costs of Poor Telecommunications

Berry (1981) presents a case study of the impact of underdeveloped communications on the French economy. His hypothesis is similar to that tested by Hardy (1980, 1981): "telecommunications are not a consequence of industrial vitality, of the standard of living; they are one of its instruments." Berry strongly disagrees with policymakers and analysts who see telecommunications as a sort of luxury arising from a certain level of prosperity. He finds that this attitude leads to underinvestment in telecommunications, which is in turn detrimental to the growth of the economy.

Berry analyzes the impact of underinvestment in telecommunications in

France, comparing France with Spain where a commitment to telecommunications investment resulted in higher telephone density, and according to Berry, contributed to Spain's substantial recent economic growth. Between 1972 and 1977, France's GNP per capita (measured in current dollars) increased by 95%, while in Spain, the increase over the same period was 145%. Spain's economic growth rate during this period was not only significantly higher than that of France, but also higher than that of the 12 largest OECD countries.

Berry's approach is somewhat similar to Parker's (1981), in that both attempt to estimate the negative impact on an economy of a telecommunications gap. However, Parker estimates the impact on the U.S. economy if there had not been a government subsidy program for rural telecommunications, while Berry estimates the impact on the French economy of a lack of sufficient investment in telecommunications.

In this study, Berry compares two European countries, France and Spain, at different levels of development and examines the growth of their telecommunications systems in the 1960s and 1970s. This comparison highlights the very different strategies followed by these two countries concerning telecommunications investment. France is a highly industrialized country which for 30 years after World War II dramatically underinvested in the development of its telecommunications, while Spain, was, until quite recently, a semi-industrialized country with a much lower level of income per capita, and deliberately followed a policy of high investment in telecommunications.

Berry finds further evidence of underinvestment in France and high investment in Spain by comparing the ratios between per capita GNP and the total number of telephones per 1,000 population. This ratio serves as an indicator of how much a country has invested in telecommunications in relation to its wealth; the lower the ratio, the higher the country's comparative investment in telecommunications. For Spain, the ratio in 1973 stood at 9.7, a figure similar to that of the most advanced countries in telecommunications (e.g., the United States with 9.7, or New Zealand with 8.6). France, in contrast, had a ratio of 22.6, indicating a comparative investment in telecommunications approximately 2.5 times lower than Spain or the United States, and about 1.5 times lower than the OECD area as a whole (Western Europe, North America, Japan, Australia, and New Zealand).

It is not clear from Berry's analysis that investment in telecommunications contributed substantially to Spain's rapid economic growth, but it appears that investment in telecommunications is likely one significant factor. Some authors (e.g., Kaul, 1978) have argued that telecommunications is most important in the tertiary sectors (such as service industries, banking, or commerce). Thus it would be important to analyze growth in the Spanish economy by primary, secondary, and tertiary sectors. It would also be important to analyze data on economic growth and telecommunication investment in France and Spain over a much longer time period, as the effects of improved telecommunications may lag several years behind investment. Also there were major political changes in Spain and economic dislocations worldwide because of increased energy costs between 1972 and 1977.

Another approach to estimating the comparative impact of telecommunications investment would be to apply Hardy's methodology to these cases in the same way that Parker (1981) applied it to the U.S. economy to estimate the economic impact of underinvestment in telecommunications.

Berry provides more convincing evidence of the cost of poor telecommunications in an analysis of what he calls the "cost of default" (*coût de défaillance* in French), namely the total of the losses to industrial and commercial organizations due to poor or nonexistent telephone facilities. We might term this the cost of the telephone gap. Berry estimates this cost in terms of:

- orders lost by industry and commerce for domestic and export business;
- losses of time by management personnel;
- less efficient use of production facilities;
- waste of energy for travel and transportation because of the lack of opportunity for timely communication;
- waste of time of service personnel, equipment, vehicles, etc. because of the lack of ability to direct them to the right place at the right time.

Based on a study of the telecommunications requirements of business and management, the French Telephone Users Association estimated the cost of the telephone gap at about 20 billion French francs, or about 2% of France's GNP in 1972.

To put this result in perspective, France invested a total of 4.1 billion francs in its telecommunications system in 1970, and 12 billion francs in 1975. In other words, the losses incurred because of poor service were at least twice as high as the total annual investments in telecommunications. Berry points out that this estimate was rather conservative, as it did not take into account other factors including:

- loss of life or goods (e.g., due to fire) when help cannot be summoned in time in an emergency;
- loss of potential jobs because plants and offices were not located in France due to inadequate telecommunications;
- less efficient use of public collective facilities (e.g., hospitals).

A study conducted in France in 1979 estimated that between 1955 and 1975, approximately 400 multinational companies established European headquarters in countries other than France (principally in London and Brussels) because of poor telecommunications in the Paris area. The losses of employment (with up to 200 employees per company unit), income, foreign exchange, and other economic advantages were substantial, according to Berry.

Since 1975, France has made an enormous effort to bring its telecommunications system up to world standards. But, as Berry observes, catching up at an accelerated pace involves heavy costs, which should be added to the "cost of default." The rapid selection, purchase, and installation of large quantities of new equipment involves major inefficiencies in the allocation of resources. The rapid hiring and training of inexperienced personnel for maintenance and operations is not always satisfactory. In addition, the equipment manufacturers which have built up their production capacity to meet new demand are faced with difficult readjustments once the most urgent needs have been met.

Berry's analysis of the French situation in the early 1970s and his comparison of France and Spain indicate that if the economic importance of telecommunications is underestimated, the cost of catching up with real demand may be much higher than the cost of the incremented development of the network.

Chapter 4

The Role of Telecommunications in Rural Economic Development and Delivery of Rural Services

4.1   Obstacles to Rural Services

Many financial obstacles impede the development of rural telecommunications services. Costs of installing and maintaining rural services are likely to be significantly higher on a per subscriber basis than costs of urban systems. The ratio of revenues to costs may be considerably lower both because of the higher costs and because of the lower population density. Since rural incomes are generally lower than those of urban residents, rural residents may be less able to afford to use telecommunications. And telecommunications authorities are likely to have ongoing pent-up demand for more profitable services such as urban connections and interurban and international circuits. Project appraisal criteria used by lenders which emphasize financial profitability and rates of return on investment more than economic and social impact also discourage rural investment.

Perhaps the "Achilles heel" of telecommunications is its potential profitability. The telecommunications sector is one of the few services provided by most governments that can be expected to pay for itself, and turn a profit as well. Profits are sometimes reinvested in telecommunications expansion or improvements of facilities; they may also subsidize other services such as the post. In some countries the revenues are simply turned over to the national treasury. Many other types of infrastructure are not expected to pay for themselves directly. Roads are constructed to facilitate growth and development; even toll roads are not expected to be profitable. Water supplies are installed to improve the quality of rural life. The provision of electrical power in rural areas is considered to be a necessary investment for development, and is generally not expected to be profitable. But, because telecommunications services are potentially financially viable, other justifications for investment in this sector tend to be overlooked.

One of the theses of this book is that the indirect benefits of telecommunications should be taken into consideration in planning and evaluation of telecommunications systems. The indirect benefits in rural areas may be particularly important to consider because demand as shown by revenues may not accurately reflect need to communicate nor benefits derived from communications.

Yet even when national planners and development institutions are willing to consider the indirect benefits of investment in telecommunications, such benefits are difficult to predict and to quantify. Similarly, planners receive little guidance in where and when to allocate their telecommunications resources for maximum developmental impact. The studies reviewed in this chapter provide many new insights into these issues.

Together these studies provide evidence on who uses the telephone in rural areas, for what purposes, and with what benefits. They also suggest that telecommunications may be particularly beneficial at certain key points in the process of rural development. The studies also provide some optimistic financial evidence. Both in India and in Egypt, rural demand in many areas greatly exceeds projections (and capacity of facilities), and rural residents are willing to spend considerably more of their incomes than their urban counterparts to communicate.

4.2   The Role of Telecommunications in Agriculture and Fisheries

The GAS-5 (Automonous Study Group 5) Committee of the International Telegraph and Telephone Consultative Committee (CCITT) has published a series of reports on the effects of telecommunications on national economies (see International Telegraph and Telephone Consultative Committee, Autonomous Study Group (GAS), *Economic Studies at the National Level in the Field of Telecommunications 1974–1976*), which have become increasingly sophisticated in their treatments of the probable effects of telecommunications on national economic development. The ITU GAS-5 manual points out that for agricultural sectors, the improved penetration of telecommunications can have organizational impacts on agricultural production through improvements in the ordering and delivery of supplies and equipment, the more timely and increased use of veterinary services and other advisory services, and the increase in the availability of current marketing information (see also Pierce and Jéquier, 1977). The results may be shown in the savings of the producer's time through more rational use of transportation facilities and through the possibility of increased specialization of production.

There is other evidence that improved telecommunications may help change the conduct of business. Interviews by Goldschmidt (1978) in Shishmareff, Teller, King Salmon, King Cove, and Unalaska, Alaska, indicated that improved, or new, telephone service, improved the ability of businesses, especially stores, to contact suppliers and customers. Supplies, in particular urgently required items such as spare parts, can be ordered on a timely basis with telephone service. Also, businesses such as guide services have found it easier to contact clients to make arrangements—this has evidently increased business.

Goldschmidt (1978) reports an interesting case of how improved communications affected the Vita packing plant in Unalaska. Due to extreme congestion on the two VHF circuits feeding into Unalaska until November 1976, telephones could not be used for frequent contacts with the rest of the world. As a result, the Vita packing plant, which packs fish and crab for shipment to Seattle, operated virtually as an autonomous plant. Its contacts with Seattle were primarily historical—what was shipped and when. Even this communication was unreliable as bad weather often hindered the shipment of the mail.

With the introduction of the satellite earth station, which virtually eliminated congestion, this condition changed. The Unalaska plant now maintains frequent contact with its Seattle headquarters (several times a day). As a result, the plant is now more responsive to customer demand—if a customer places an order in Seattle, this can be conveyed instantly to Unalaska and acted upon. In the past, shipments were made to Seattle in the hope that they would match demand. Also, if prices in New York for a particular product, e.g. Tanner Crab, increase, then Unalaska can change its fishing operations to take advantage of this information. This has resulted in improved efficiency at the plant, better customer relations as specific orders are filled, and a shift in managerial control making the Unalaska plant a satellite, rather than an autonomous plant as in the past.

A study by Dubret (1981) indicates how the fishing industry has in recent years been revolutionized by the advent of sophisticated telecommunications equipment. A main concern of fishermen is to minimize fuel consumption in finding shoals of commercial fish. Dubret points out that the importance of telecommunications can be gauged by the fact that fishing fleets need one liter of fuel to catch from three to six kilograms of fish. With world demand for fish increasing at the rate of 10 to 20% per year, and the price of oil exceeding $30 per barrel, savings on fuel become imperative. Dubret reports that a fishing fleet of 7 vessels in the Mediterranean can catch up to half a ton of tuna per day using radio communications to signal to the fleet the location of a shoal of fish.

Dubret also describes some of the electronic communications equipment which has increased the harvest and profitability of modern fisheries. These technologies include echosounders and sonar to detect fish, the ichthyoscope which identifies commercially important fish, and the trawl sounder which monitors the nets. Radio communications for navigation as well as for information on weather, tides, and water conditions are also essential to the modern fisherman.

Dubret points out that telecommunications can be important to marketing of fish. Buyers must be found for fresh fish within hours, to avoid losses from spoilage. The international marketing agency INFOPESCA, based in Panama City, serves Latin American and Caribbean fish importers and exporters in 22 countries. In 1978, this service resulted in sales of more than 100,000 tons of fish worth $100 million.

Dubret suggests that use of telecommunications equipment to support fisheries can help to increase the world's food supply, and that efficient marketing can help make fish affordable for the world's poor. In addition, commercial fishing can be an important source of revenue for developing nations. Benefits of these technologies for fisheries might be measured using techniques similar to those applied by Hudson (1981) for agricultural production and marketing. However, a quantitative approach might underestimate the importance of telecommunications, because without such telecommunications facilities, the modern fishing industry as we know it today could not exist. Conversely, the traditional fishing fleets of the poorer countries which still operate without any modern detection and communications equipment, find themselves at a considerable disadvantage. Their familiar fishing grounds are being systematically exploited by sophisticated modern trawlers from other countries, and the survival prospects for these technologically outdated fleets appear dim. Investments in better telecommunications facilities may be a necessary step for their long-term survival.

A statistical analysis performed by Wellenius, Castillo, and Melnick (1971) found that it was possible to rank different types of agricultural productive units according to the probable relative effect of the telephone service on the efficiency of production. The authors found that:

> at least when the number of different types of productive processes is small, one can quite readily derive qualitative models of the productive function that reflect clearly the main external transactions (flows of input goods output goods technical assistance, travel, administrative operations, financial operations, etc.). Then each of the transaction flow lines can be analyzed to determine, at least subjectively, the likely effect telephone service would have on these if available. A process of synthesis leads to a ranking by telephone effect on production. It was also found that to characterize the production function (from the information-flow point of view) the main factors are rather few and they can be identified in practice. (Wellenius, 1972)

Several Russian studies have also examined the role of communication in production, particularly in agriculture and large industries.[1] Voronov (1969) studied the consumption of communication output, and described the procedure for using the findings of the investigations in drawing up intersector input/output balances. He points out that early studies examined only that part of tariff revenues associated with made up consumption of communication output by the material production sectors, ignoring consumption for services and public uses

---

[1] These studies in English translation can be found in "Soviet Studies on the Indirect Benefits of Investment in Telecommunications" (mimeo), OECD Development Centre, Paris, 1978 (Working Paper 5 of the Expert Meeting on Methodology for the Study of Socio-Economic Benefits of Telecommunications).

which were considered nonproductive. Voronov concludes that the attribution of a considerable part of communication output to services and corresponding activity of communications people to nonproductive work resulted in curbing the rate of capital investment in communication development, resulting in a lag in development of communication facilities, particularly local telephone networks. Citing evidence from the Soviet Union, Voronov states that during the first five-year plan, telecommunications accounted for 1.7% of total investment, but the share fell to 1.0% during the second plan, and to 0.53% in the fifth plan.

The author concludes that, although before the period of rapid techonological progress, communication development was primarily prompted by private demand, in the period of expanded automation of production and control the increased demand for communication output now comes from the production sectors. The interbranch material balance (or input/output sector analysis) can be a source of data for estimating the consumption levels of various sectors. The author makes several suggestions for improving the data collection and analysis used to document the utilization of communications output.

Mednikov (1975) studied the role of dispatcher communication (mobile radio) for operational control of agricultural production designed to increase output volume and labour productivity. The author states that dispatcher communications ensures a 2- or 3-fold reduction in the idle time of machines and tractors which raises the productivity of the machine and tractor fleet; a cut in performance times for basic agricultural operations, a rise in cattle productivity in all production subdivisions as a result of constant checking of observation of animal husbandry standards regarding the feeding and care of livestock; a reduction in the time spent by leaders and specialists on operational control of production; a cut in the time needed for various kinds of removal and transfer and considerably improved coordination of farm technology services.

Mednikov then describes a dispatcher communication system in the Rostov region which included links to 60 collective and state farms and to the regional agricultural data processing center. The improved operational control of production resulting from the regional/district/farm communication system was predicted to contribute to increased yield of grain and milk. However, no control areas were studied.

4.3   Rural Telecommunications and Development in India

A study by Kaul (1981) examines usage of two types of telephone service available in rural India, namely the subscriber telephone leased to individuals and the public village telephone ("long distance public telephone" or LDPT). This study raises the question of trade offs between private and public services. Given limited resources, should a telecommunications authority allocate more of its resources to the potentially more profitable subscriber services or the public telephone services? Central to this question is an analysis of benefits to determine if other factors than revenue generation should be considered by national planners.

Despite having much more extensive telecommunications facilities than many developing countries, rural telephone density in India is very low. The rural areas have an average telephone density of 0.03 per 100 population, which is almost 60 times lower than the urban areas' density of 1.71 per 100 population and 12 times lower than the national average. Of India's 2.01 million telephone lines in 1980, only 140,000 (or 6.9% of the total) were located in the 580,000 villages which account for 80% of the country's total population.

India's telecommunications plans for this decade call for major expansion of the national network. By 1990, the total number of telephones should exceed 7 million lines, i.e., 3.5 times more lines than 10 years earlier. However, this growth is unlikely to meet the anticipated demand for telephone services. Thus, one of the crucial issues facing telecommunications planners is the distribution of the additional exchange capacity to be installed during the 1980s. Planners must decide what mix of urban and rural facilities to install, and how to distribute the facilities in rural areas to achieve maximum developmental benefits. Kaul's research provides important insights on these issues. However, he does not address what role (if any) India's domestic satellite (INSAT) should play in provision of rural telecommunications.

Kaul (1981) found that of rural subscribers (i.e., who lease their own telephone), 66% were employed in business, while 20% worked in agriculture. Of users of the long distance public telephone (LDPT), 41% worked in agriculture, while 33% were in business. Although incomes of the subscribers were higher, they were far from affluent. Most of the long distance calls (63%) terminated in the district, while 30% went to other districts in the state. Only 7% of calls went outside the state.

Kaul and his colleagues also attempted to measure the effective utility of this service to the users. Their procedure was to interview a large sample of village telephone users about the purpose and urgency of the last long distance telephone call they made.

This approach of focusing on the last episode has been demonstrated in field studies to produce more reliable results than asking respondents to generalize about their behavior.

Over 75% of the users stated that their last call was of an urgent nature. Nearly 90% of those who considered their last call urgent would have been prepared to travel personally to convey their message. Using data from their respondents about the destination of their last call, Kaul and his colleagues estimated the consumer surplus of these calls, i.e., the potential benefit represented by the difference between the cost of the call and the total cost of traveling personally to convey the message. The results are shown in Table 4.1.

Table 4.1
Estimated Consumer Surplus of Long Distance Rural Telephone Calls[a]

| Called Distance (km) | Average Distance (km) | Expenditure on Telephone Call (Rs) | Expenditure on Travel by Bus | | Minimum Opportunity Cost of Person-Days Lost (Rs)[d] | Total (4+5+6) (Rs) | Consumer Surplus (7–3) (Rs) |
			Return Bus Fare (Rs)	User Cost[c] (Rs)			
Up to 20	11.24	1.37	1.88	2.65	2.00	6.53	5.16
20–50	34.57	3.54	5.80	2.65	4.00	12.45	8.91
50–100	80.54	4.56	13.54	2.65	8.00	24.19	19.63
Above 100	149.00	5.44	25.04	2.65	8.00	35.69	30.25

[a]Source: Kaul, 1981

[b]Rs = Rupees

[c]The user cost is the cost incurred by the traveler to go to and return from the bus station at the points of departure and arrival.

[d]The opportunity cost of person-days lost has been computed on the basis of the minimum wage (Rs 8 per day).

The estimates in Table 4.1 are rather conservative. The cost of person-days lost in travel was computed on the basis of the minimum wage, rather than the basis of the effective earnings of the respondents, which are generally higher. Furthermore, estimates of travel time are probably optimistic and underestimate the actual time required to make a journey by bus.

This analysis shows that the consumer surplus accruing from a telephone call compared to the price of the call is substantial. This benefit is at least 4 or 5 times higher than the cost, perhaps as much as 10 times the cost of a telephone call, given the actual travel time required. As can be seen from Table 4.1, the benefit of telephone use tends to increase with distance. While this result may appear self-evident, it is important to note the corollary that the regions which stand the most to gain from the telephone are precisely the remote rural regions which are generally poorly served by both transportation and telecommunication systems.

The Kaul study shows that in India the cost of using the public telephone is several times lower than the benefits accruing to the user. Yet at the same time, as shown by Kaul in the case of Andhra Pradesh and by others in different parts of the world, rural telecommunications tend to be a money-losing operation. Kaul raises the question of whether the consumer surplus approach to measuring the benefits of rural telecommunications might be a better basis for measuring the overall profitability of rural telecommunications investments.

Kaul suggests that developmental benefits from rural telecommunications are likely to result once a certain take-off point of economic growth has been achieved. He finds that demand for rural telecommunications increases substantially once the process of rural modernization has begun. Regions of India where telecommunications facilities are generating substantial traffic and breaking even or making a profit tend to be areas with more modern agricultural sectors.

Regions where telecommunications facilities are underutilized tend to have subsistence economies. Kaul suggests that planners should take the varying rates of agricultural modernization in India into account in planning rural telephone service.

The Economics Study Cell projects that a total of only 45,000 "access points" or long distance public telephones would be sufficient to provide basic service to all of India's rural population. In 1980, India had some 13,400 access points of this type, and the 1980–1985 plan calls for an additional 20,000 points. Both these figures represent approximately 10% of the total number of telephones in the rural areas; the other 90% are accounted for by private subscriber telephones. The figures suggest that the provision of telephone service to the rural areas is far from impossible.

The interviews with rural telephone users emphasize a number of common problems concerning rural telephone services. Access to a public village telephone is often difficult; this telephone, usually located in the local post office, is generally not accessible for more than 3 to 5 hours a day. The result is inconvenience to the user, but also reduced revenues for the telecommunications authority, which in turn makes investment in rural facilities even less attractive. This problem of access is also important because of a tariff structure which offers low rates during off-peak hours.

A second problem is reliability of service. Nearly 55% of rural users surveyed were dissatisfied with the quality of telephone service. As in many other developing countries, rural telecommunications in India suffer from poor reliability, high maintenance costs and low transmission quality. Customers are very sensitive to this problem. The major cause of dissatisfaction was low completion rates for calls, especially since many callers travel considerable distances to use a phone.

A third important issue concerns price. Village public telephones are used for long distance calls, which are priced according to the duration and distance of the call. Although Indian rates are low by international standards, the price of a long distance call is relatively high, given rural income levels. However, less than half the users (43%) felt that the tariff charged was high, and more than 11% of the users who found the tariff high would have been satisfied with a 25% reduction. People living in the rural areas are prepared to pay for telephone services, and demand for rural telephone services is apparently quite inelastic in terms of price for service, as there is no practical alternative to using the telephone.

4.4   Benefits of Rural Telecommunications in Egypt

The analysis of the benefits of rural telephone service in Egypt by Kamal (1981) and his colleagues has a number of features in common with the Indian study described above. Both studies focus on the microeconomic level—the village and the individual telephone user—and both attempt to measure these benefits through an indirect rather than a direct approach.

The analysis carried out by Kamal is based on interviews with almost 2,000 villagers in a sample of 146 villages (out of almost 4,200 Egyptian villages) located in 7 agricultural provinces. Contrary to Kaul's study in India, the Egyptian study focuses on key officials and tradesmen in the villages, rather than individual peasant farmers. The focus on key individuals in the villages was based on a previous intensive study of nine villages that revealed that the main telephone users in rural Egypt are in nonagricultural sectors. Users were clustered in four main groups for analysis: service organizations; the trade sector; owners of capital equipment, large proprietors and liberal professions; and the artisan sector.

Kamal (1981) and his colleagues break benefits from telephone usage into four main components:

- direct monetary benefits measured by the difference between the cost of the telephone call and of an alternative means of communication;
- time savings as a result of telephone use measured by the number of working hours saved;
- indirect monetary benefits measured by the value of losses avoided due to use of the telephone in emergencies;
- indirect monetary benefits measured by the efficient use of capital and capital equipment (using wages of operators to assign value).

The Cairo University team analyzed the results by sector and by village, aggregating the villages into provinces.

Table 4.2 shows a summary of the benefits for each sector summed over all villages.

These results bear striking testimony to the value of rural telecommunications. As was the case in India, urgent communications predominate, with emer-

Table 4.2
Summary of Benefits of Rural Telephone Use in Rural Egypt[a]

Sector	Total Benefits (%)	Benefit/Cost Ratio
Service organizations	59.0	85:1
Trade sector	14.3	69:1
Owners of capital equipment, large proprietors, liberal professions	20.8	126:1
Artisan sector	5.9	78:1
	100.0	

[a]Source: Kamal, 1981

gency communication accounting for a major portion of the benefits of the service sector. Benefit ratios varied greatly from province to province depending on their distance from major cities, the distance between villages and regional centers (the markaz), and the economic composition of the villages.

Kamal found that telecommunications benefits are correlated with the population of the village and telephone availability (number of direct lines to the district exchange, lines to the village switchboard, and public call offices). Benefits were also correlated with education levels; the higher the education level, the greater the benefits derived. However, occupation may be a better predictor of benefits, as the group consisting of merchants, artisans, and proprietors derives greater benefits in some categories such as efficient use of capital equipment than do more highly educated professionals and administrators. The analysis also shows that educational level correlates with administrative level of communication. Illiterates communicate primarily with neighboring villages and the markaz, while the more highly educated communicate with provincial capitals and major cities.

The major conclusions of the Kamal study are:

- Rural telecommunications services offer a potentially high ratio of indirect benefits to costs. This ratio for 146 Egyptian villages averaged 85:1. Kamal predicts that better quality of rural telephone services would result in higher utilization and higher rates of return.
- Total population of the village is the best predictor of telephone benefits, particularly for the service organization sector.
- Availability of adequate telephone service (indicated by number of exchange lines and PCOs) is an important factor in promoting the use of telephone service and increasing the level of benefits derived.
- Distance from major centers also correlates highly with telephone benefits. The greater the distance to the markaz (district center) and the provincial capital, the greater the benefits. Distance to major cities is particularly important for large proprietors and owners of capital equipment.
- Education is correlated with benefits from telecommunications use in each occupational category; the higher the education level, the greater the benefits derived.
- In Egyptian villages, the service organizations derive greatest benefits from telecommunications followed by large proprietors, the trade sector, and artisans.
- Egyptian villages are extremely diverse in the factors that could influence telecommunications use and benefits. However, analysis at the provincial level revealed meaningful patterns and correlations that will be further explored in additional research.

4.5   Communications to Support Agricultural Marketing: The Cook Islands Two-Way Radio Network

The communications difficulties in the Cook Islands in the South Pacific stem from the fact that the country's population of 18,000 lives on 13 islands scattered over 2 million sq. km. of ocean. Eight of these islands have airstrips, but for the others the only means of communications is the ship, which may call no more than 6 to 8 times a year. More than 54% of the country's export income in 1978 came from canned fruit juice and canned pineapple, but a greater emphasis is now placed on the export of fresh fruit and vegetables, which bring much higher prices, but which are also much more dependent on rapid transportation and good telecommunications. Good transportation and telecommunications are therefore vitally important for development of agricultural exports.

The effects of better transportation facilities can be quite dramatic. Following the opening of an international airport near the capital, Rarotonga, in 1974, the exports of fresh fruit and vegetables leaped from less than 50,000 kg. in 1974 to 400,000 kg. in 1980. With a distance of some 3,000 km. between the Cook Islands and New Zealand, its main export market, airfreight is the only means to deliver perishable produce to market. However, good transportation must be coupled with supporting telecommunications in the marketing of fresh produce. Farmers on the various islands must be able to notify shippers when they have fruit or vegetables ready for export. Conversely, the farmers must know when to expect the ships (which transport the produce to Rarotonga for export) so that the produce is ready for shipping, and yet not harvested too soon to risk spoilage.

A high frequency, two-way radio network for government services has been in operation since 1979. It consists of a base station in Rarotonga, two additional stations for the Premier's use in Rarotonga, and stations on 12 other islands. The operator at the base station stands by to pass messages four times per day, patches island calls into the local telephone system, and follows up on information and action requests. The remote stations are operated by government administrative officers, or where this is not practicable, by part time operators.

In addition to point-to-point messages, the system is used for regularly scheduled conferences for officials in the ministry of agriculture, the shipping industry, and the health ministry, among others.

The system carries an average of 500 messages per month, or over 6,000 messages per year. Of 18 institutional users, major users during a sample period were the ministry of agriculture, the agricultural marketing board, and the government departments of internal and outer islands affairs.

The capital cost of the system is roughly $30,000, with annual operating costs of $12,500. Assuming amortization of capital costs over 5 years, yearly costs would average $18,500. With an average of 6,000 calls per year, the cost per call would be about $3.00. One of the most important uses of the system is to coordinate the shipping of agricultural produce to markets. Every Monday morning a representative of the agricultural marketing board speaks by radio to agricultural offices on each of the southern islands to determine how much perishable produce they will have ready for shipment that week. Then on Tuesday morning the shipping committee (comprised of representatives from the shipping company, the cannery, the marketing board, the ministry of agriculture, and other government officials) meets to decide where the ships will be sent that week. This information is then passed to agricultural officials on the islands by radio so that they will be able to alert the farmers when to expect the ship.

Previously, to prevent spoilage, farmers would not harvest their crop until the ship was on the horizon. The result was small shipments and longer layovers by the ships while waiting to load. Farmers who misjudged the arrival time of ships risked spoilage of their produce. Agricultural officials state that the use of the radio system to coordinate shipping has significantly reduced spoilage. For example, to save crop wastage, the radio can be used to redirect a ship by contacting its next port of call. In this way 13 tons of pineapple worth $1,950 were saved as a result of one call to redirect a ship, costing $3.00.

Another effect of the radio system is to improve efficiency of the shipping operation. The cost of each shipping call to a southern island is estimated at $2,830, while shipping calls to northern islands cost $3,400 each. The total annual cost of interisland shipping service in 1978–1979 was estimated at $515,000. Thus the use of the radio to improve shipping efficiency can increase the benefits of this expensive service.

Finally, the use of the radio to coordinate shipping can result in higher prices for agricultural produce. There is considerable economic importance in assuring that produce is picked and shipped at the optimal time to reach the export market in prime condition. Lower quality produce brings much lower prices. As the Cook Islands attempts to diversify its agricultural base and to export both fresh fruit and vegetables, the importance of a coordinated communication and transportation system will increase in significance.

It should be noted that the radio network is used for other government activities to support the agricultural sector. For example, one island has increased its export of fresh vegetables through improved production practices. Extension workers have taught farmers new techniques, with support from the ministry of agriculture in Rarotonga, which can now act as a resource for the whole country using the radio. The secretary of agriculture stated that it is now easier to get agricultural staff to work in the outer islands because of improved communications. Letters take from three weeks to three months to reach them. Radio telegrams could be sent in emergencies, but were not considered a satisfactory means of supporting isolated field staff. Finally, agricultural officials are able to get expert assistance from other institutions in the region, including the University of the South Pacific and the Univeristy of Hawaii, through a conferencing link over the ATS-1 satellite. (This system is described in more detail in section 4.8.)

Hudson (1981) points out that there are several external constraints which limit opportunities to maximize full benefits from the radio network. Among them are the fact that there are bottlenecks in the transportation system because of inadequate shipping capacity. The Cook Islands also face competition in export markets from other countries in Southeast Asia and Latin America with cheaper labor costs or more efficient plantation-style agriculture. Most agriculture in the Cook Islands is undertaken by farmers with small individual holdings.

External constraints are likely to be found in many developing countries. However, the experience in the Cook Islands indicates the extent to which telecommunications when combined with other essential infrastructure, in this case transportation, can make a significant contribution to the growth of the agricultural sector.

4.6   Rural Telecommunications in Africa

Clarke and Laufenberg (1981) undertook a comprehensive examination of existing telecommunications facilities and requirements in Sub-Saharan Africa, emphasizing rural regions. The authors analyzed available data on telecommunications and economic development in Africa, and carried out numerous micro studies of applications of various telecommunications facilities to support development activities. They selected the following nations for special attention: Botswana, Ivory Coast, Kenya, Niger, Senegal, Tanzania, Upper Volta, and Zimbabwe.

Many of the nations on Sub-Saharan Africa are among the least developed in the world. Twenty African nations are cited by the United Nations as among the world's least developed, while Africa accounts for 25 of the 38 countries considered by the World Bank as low-income nations (GNP per capita less than U.S. $360 per year).

Africa remains a rural-dependent continent, with approximately 75 to 80% of its population living in rural areas or deriving their livelihood from rural economic activities. Land areas are vast, with great variations in regional population densities. For example, the rural regions of the Sahel nations in West Africa are much more sparsely populated than rural Nigeria or Zimbabwe. Telephone density and investment in telecommunications in Africa are strikingly low, even compared with other developing regions.

Aside from South Africa, which has about half of the continent's telephones, in 1978 Africa accounted for only 0.5% of the world's telephones, but 10% of the world's population. Network expansion has been relatively slow; between

1970 and 1978, the total number of telephones in the African continent aside from South Africa has increased from 1.65 million to 1.91 million, an annual growth rate of only 1.65%. The existing systems are heavily concentrated in the urban areas; of the 28 countries for which data are available, 8 had more than 90% of all their telephones in principal reporting cities, while in another 8 countries, between 80 and 90% of total telephone capacity was in the urban areas. Vast rural areas have no telecommunications facilities at all. Rural areas with telephone service usually have more limited access to telecommunications than the data would suggest. Quality of service is generally much lower than in urban areas, waiting lists are longer, and the equipment often consists of out-dated exchanges recycled from the urban network.

Clarke and Laufenberg (1981) examine the potential benefits of increase investment in telecommunications in Africa, especially in rural areas. The au-thors shed some additional light on the relationship between telecommunications investment and economic development. They suggest that this relationship prob-ably follows a step-like function: the strong relationships between telephone density and income per capita are clustered at various points along the income range, indicating that telecommunications have a very strong multiplier effect on economic growth at certain specific stages of development. In other words, telecommunications investments may trigger economic growth at certain times. Conversely, investments in telecommunications at periods other than those of sudden take-off may appear not to affect economic development. One implica-tion of this observation is that the development effects of telecommunications build up over time, and become dramatically visible, for limited periods, when a critical mass of telecommunications investment and economic activity has been reached. Clarke and Laufenberg hypothesize a step function that may occur at several stages in the development process.

Another important macroeconomic aspect of telecommunications is the link-age with international trade. An analysis of international trade and international telecommunications traffic for six Sub-Saharan African countries shows a very close association between outgoing international calls and per capita GNP ($R = 0.95$). The correlation between international calls and export trade is a little weaker ($R = 0.86$), but this correlation is comparatively stronger with the ex-ports of food and agricultural raw materials ($R = 0.89$), and even stronger if ores and metals are added to agricultural exports ($R = 0.90$). It appears that good telecommunications is an important instrument in the development of exports from the rural zones. This finding appears to contradict the notion that telecom-munications are important mainly to the export of industrial products.

Clarke and Laufenberg provide examples of the potential benefits of rural telecommunications investment in rural Africa by citing ITU cost/benefit studies carried out in 1976. The projects analyzed were primarily to support agricultural activities in Ethiopia, Lesotho, Malawi, Somalia, and the Sudan. Conservatively calculated benefit-to-cost ratios ranged from 2.08/1 to 8.28/1. The authors cite numerous other examples of rural agriculture, resource extraction, and fisheries, where an investment in telecommunications would appear to be highly cost-effective.

Clarke and Laufenberg (1981) suggest that systematic provision of Public Call Offices (PCOs) may be the best approach to increasing telephone access in rural Africa. They suggest a goal of one PCO per 100 km^2, or a minimum of one PCO per 10,000 population. (Note that this was the density used by Hudson et al. (1981) for their hypothetical rural region.) At least 40,000 new PCOs would be needed in Africa by 1987 to meet this minimal target. They suggest that a satellite system optimally designed for rural services may be the best way to meet these requirements. Small satellite earth stations could serve clusters of PCOs linked by VHF or UHF radio.

In their analysis of Sub-Saharan Africa, Clarke and Laufenberg do not suggest that if more and better telecommunications facilities were available to the rural areas or the lower income groups, total income would be better distributed. Rather, they conclude that if rural areas are unlikely ever to get a minimum amount of telecommunications service, then they are even more unlikely to derive any benefits from this nonexistent service. Since there will be investment in telecommunications, the benefits of this investment will accrue to those regions which get service, which are generally the urban areas. Thus urban/rural income disparities will increase. Conversely, carefully planned investment in rural telecommunication may have the opposite effect and may foster the economic growth of rural agricultural as well as industrial sectors.

4.7   Rural Delivery of Education and Social Services

Most documented experience in applications of communications for education has focused on the mass media. Radio broadcasting has been used to bring quality instruction to rural primary school children who have difficulty in learning because of overcrowded classrooms, poorly trained teachers, and lack of learning materials. Radio has also been integrated in many rural development campaigns designed to increase adult literacy, provide practical skills training, improve maternal and child health, etc. The worldwide proliferation of transistor radios plus mixed results of educational television projects have stimulated a renewed interest in radio as a less expensive but potentially highly effective education medium.

Two-way communication has been used to support correspondence studies in many developed countries. In Australia, the "School of the Air" enables children on remote sheep stations to talk to their teacher and to "classmates" as part of their correspondence studies. Educators in the United Kingdom, the United States, and Canada are using a variety of interactive technologies including audio and computer teleconferencing, audio-graphics and facsimile to communicate with distant students. As with health services, telecommunications can also be used for administration to improve the efficiency of educational services.

Early experiments using the ATS-1 satellite for education in Alaskan villages demonstrated the role of telecommunications in providing administrative and logistical support:

> In the past, generator failures, furnace blow-outs, or deficiencies in instructional materials would cause the whole school to shut down while a written message was mailed out. With the satellite radio, teaching conditions were greatly improved and teachers would call for immediate emergency parts during the critical winter months. The communications links between the schools and the area administration supported and encouraged the efficient operations of the school. (Parker, 1973)

More recently, Alaska has installed an electronic mail system with links via satellite to computer terminals in each of its rural school districts (see below).

A World Bank project addressed the need for better telecommunications to support educational reform in the Philippines. New textbooks had been published, but without reliable communication links, the government could not ascertain that the books had in fact been distributed from regional warehouses to schools throughout the country. A pilot project involving installation of thin route satellite stations in rural regions of the Philippines will provide an opportunity to assess the benefits of better communications to support education. Indonesia's PALAPA satellite will be used for the project. Another project using PALAPA sponsored by the U.S. Agency for International Development will link universities throughout Indonesia for audio, graphic, and data communication.

Several studies and projects have indicated the value of telecommunications for social service delivery as a means of extending the range of the institutions, providing training and consultative support to local minimally trained field personnel (e.g., health workers and teachers' aides), and improving the administration of field activities. A major difficulty in most rural development and social service delivery projects is management and coordination. Given the training deficiencies of rural workers, communication links can be essential to providing supervision and continuing education (e.g., through supervised on-the-job training). The costs of such management supervision are likely to be prohibitive if extensive travel by professionals or middle-level managers is required. Regular voice communication can maintain an effective network for management supervision and continuing education.

Audio teleconferencing via satellite has been used for more than a decade to link village health aides in Alaska to regional hospitals (see below). In other parts of the world HF and VHF radio are used by paraprofessional health workers and nurses to communicate with their supervisors for medical consultation, administration, patient follow-up, and continuing education.

Some rural health services use communications primarily to coordinate transportation. In particular, flying doctor services in Australia and Africa use HF radio to coordinate their aircraft and field personnel. Telecommunications has also been used to provide critical assistance in natural disasters, including earthquakes (e.g., Nicaragua), floods and tidal waves (the Pacific islands), volcanic eruptions (the United States), and outbreaks of epidemics such as cholera and dengue fever (the South Pacific).

Studies on the benefits of telecommunications tend to focus on the day-to-day uses of the telephone. Yet in an emergency, the immediate availability of a telephone can made the difference between life and death of people or survival and ruin of their communities. Emergency communications appear to be a very important function of rural telephone service worldwide. Data collected by Kaul (1981), Hudson (1981), and the World Bank (1976) all show emergency communications accounting for 3 to 5% of rural telecommunications traffic. However, as Kamal (1981) shows, the benefits from such communications may be proportionately much higher. His analysis of the service sector shows that more than 30% of the total benefits accruing from the use of the telephone in this sector result from emergency use. This is not to say that 30% of telephone calls are made for emergencies, but that the benefits are comparatively much larger than for other types of calls. Kaul's investigation of the purposes of the last telephone call shows that emergencies and health reasons account for 8.6% of the total number of calls.

4.8   Telecommunications to Support Distance Learning: The University of the South Pacific's Satellite Network

Hudson (1981) conducted a study of the cost effectiveness of audio conferencing via satellite to support distance education in the South Pacific. The University of the South Pacific (USP) was established in 1967 to serve the island nations of the South Pacific. The islands are scattered over an area as large as Australia or the continental United States, but their total land area is very limited, and their total population is only about 1.5 million. Most nations are newly independent, while a few are currently semi-autonomous with their own internal self government. The three socio-geographic areas of the island Pacific are represented, including Polynesia in the north and east (Western Samoa, Tuvalu, Tokelau, Niue, Tonga, Cook Islands), Micronesia in the northwest (Kiribati, Nauru), and Melanesia in the west (Fiji, Solomon Islands, Vanuatu).

The University has a mandate not only to serve the higher education needs of this diversified region but also to act as a source of developmental expertise. To help meet the development needs of the region, USP has established an extension services program including in-country university extension centers. Each center provides a link between the university and the country: it coordinates the administration of correspondence courses; it provides guidance to potential students about university options; it acts as an information retrieval and resource center for governments and individuals seeking development information; and it acts as a local resource center for educational and cultural activities. Each of the centers is linked to the others and to the main campus in Fiji and the College of Agriculture in Western Samoa, by NASA's experimental ATS-1[1] satellite.

Each site is equipped for audio conferencing via the satellite. Sites are also equipped with audio cassette recorders and microcomputers for transmission of hardcopy messages as a form of electronic mail system. Three of the sites are equipped with slowscan video equipment which can transmit a still television picture over a narrowband voice channel.

Transmission time is made available free by NASA. Assuming that capital costs are amortized over seven years, the cost of the system including capital and operation is approximately US $112 per hour. Assuming an average of 15 participants per session, the cost per participant hour is approximately $7.50.[2]

USP uses approximately 23 hours per week of satellite time, primarily for administration, course tutorials, and outreach programming. Schedule coordination is no small task, as the university's extension centers lie on both sides of the international dateline, and in four time zones.

The administrative use of the satellite network for staff meetings, logistics, and course planning is the mainstay of the system. At a cost of US $112 per hour, it appears to be a particularly effective means of managing a regional institution. For meetings, the only alternative is travel to a common destination such as Suva. The estimated cost of a meeting involving one participant from each USP site is $8800 based on travel costs and time away from their work. An 8-hour satellite meeting would cost only about $900, for a benefit-to-cost ratio of about 10 to 1. In practice, the length of such meetings is much shorter, so that the actual benefit-to-cost ratios are likely to be higher.

For routine administrative matters, the major alternative is the mail-bag system. Although much improved over early years, it still requires a minimum of a week turnaround to get a response in writing from Suva. This system is an important component of extension education management, but total reliance on the mail would greatly decrease the efficiency of regional extension services.

Time devoted to satellite tutorials has increased by 375% since 1975. It appears that the skill of the instructor is the key factor in satellite tutorials. An

[1] ATS-1, launched in 1966, was the first of the Applied Technology Satellites developed by the U.S. National Aeronautics and Space Administration (NASA).

[2] No commercial audio conferencing service is currently available in the region. However, such a service could be provided using the countries' international INTELSAT stations. An annual tariff of US $123,000 was proposed for 1000 hours of conferencing time. Including staff costs, Hudson (1981) estimates such a commercial network would cost USP $173 per hour, or $11.50 per participant hour.

alternative format to direct student tutorials is a session involving the instructor in Suva and local tutors at the centers. The tutors work directly with the students, and then present the problems they are encountering to the instructor, and follow up personally with the students. Extension center directors generally felt that local tutorial assistance was the most important factor in successful completion or courses. Thus, the satellite tutorial may be considered a contributing factor when coupled with an effective local tutor and when conducted in an effective manner.

An in-depth analysis was made of three courses with particularly effective satellite tutorials. Completion rates of students with access to tutorials (46.2%) were almost double those of students without access to tutorials (23.3%). The potential benefits to USP of effective satellite tutorials are substantial. At a cost of $7.50 per participant hour, for 15 weeks the cost per student per course may range to $117.50. However, set against a total cost per student per course estimated at $440 to $550, if the investment in tutorials results in an improvement of the completion rate, the benefits can well justify the costs.

The term "outreach" refers to satellite applications that are designed to apply the resources of the university to development problems in the region through consultations, seminars, and/or specially designed and targeted short courses; and to uses by other development agencies serving the region. Also included in this category could be USP participation in the PEACESAT[3] network coordinated by the University of Hawaii. USP's School of Agriculture in Western Samoa has used the satellite network for consultation with experts (e.g., agronomists, soils scientists, veterinarians), short-term training courses, follow-up to on-campus courses, and regional administration.

The USP satellite network has been used for several types of health applications by the World Health Organization and by physicians in the region. These include:

- consultation to remote physicians,
- emergency support for disaster relief and epidemics,
- in-service seminars, and
- regional administration.

The USP satellite network was used to coordinate assistance for an outbreak of cholera in Kiribati (formerly the Gilbert Islands) in 1978. A conference link was established between Tarawa (Kiribati), Suva (Fiji), Wellington (New Zealand), and Noumea (New Caledonia), to coordinate the emergency airlift of drugs and supplies and to provide guidance on treatment and quarantine procedures. The system has also been used to assist in coordinating relief activities

[3] PEACESAT is an acronym for Pan Pacific Education and Communication Experiments by Satellite.

for outbreaks of dengue fever and Ross River fever in several countries, and to alert medical staff in other nations in the region of these outbreaks.

Benefits from outreach activities are the most difficult to quantify. Some benefits may be direct and dramatic, if contact by satellite can result in a consultation or relief activities that are lifesaving. Similarly, the benefit of a consultation that eliminates the need to evacuate a patient will save thousands of dollars in transportation and hospitalization costs. And a seminar that disseminates important agricultural information to agricultural extension officers can replace a week-long seminar in Western Samoa, with benefit-to-cost ratios in excess of 10 to 1 in direct travel savings alone.

Agricultural and other developmental applications are likely to have longer term benefits. Seminars which transmit information that can improve crop yield may eventually contribute to an increase in agricultural output and potentially greater foreign exchange earnings. Further use of correspondence courses with satellite tutorials could enable agriculture students to take part of their degree program in their home countries, thus saving USP at least $3,000 per student per year.

For many other applications, no practicable alternative to the use of the communication system is available, so that the satellite network produces benefits that would not be possible without the system.

4.9   Satellite Communications to Support Rural Services in Alaska

The population of Alaska is about 400,000, of whom about 60,000 are Indians, Eskimos, and Aleuts. Most of this native population lives in widely scattered villages of 25 to 500 people. The average distance between one settlement and another is approximately 60 miles; some villages are more than 150 miles from their nearest neighbor. Travel between remote communities is primarily by bush plane. Storms, extreme cold, and rough, unlighted airstrips make flying hazardous and at times impossible. Difficulties in transportation make communications extremely important.

In 1971, Alaska gained the opportunity to see if satellite communication would help to improve village health care. Twenty-six sites in Alaska were chosen to participate in medical experiments using NASA's ATS-1 satellite. In 1975, the Alaska Legislature provided funds for the purchase of 120 small earth stations to provide telephone service to all communities with a permanent population of 25 or more, using the U.S. domestic RCA SATCOM satellite system.

Small village earth stations (with 4.5 meter antennas) are usually equipped with two circuits: a conventional message telephone circuit, and a special circuit for medical service. Expansion up to eight circuits can easily be accomplished through the insertion of additional circuit modules to supply enough trunks for a community of several hundred with a local telephone exchange. In villages without a local exchange a public telephone is located in a central building accessible to the public, and is generally monitored by an attendant. A special handset is located in the clinic or health aide's home.

Primary health care in the villages is generally provided by native health aides who receive basic training from the Public Health Service. The health aide returns to the village with basic tools: a drug kit, instruments, reference manual, and a satellite communication link to the regional hospital.

The satellite network design is based on experience with HF radio and later with ATS-1, where health aides and the regional hospital shared a single audio channel. It was found that the shared channel had the advantage of enabling the health aides to listen to consultations between the physicians and other aides, thus improving and refreshing their knowledge. Listening in to other medical traffic also reduced the sense of isolation. The abilities of the patient to hear the doctor helped to reinforce the health aide's advice. In addition, there was a potential for using the system for training in a "broadcast" mode. These features were combined in the operational medical network. Two hospitals and their associated villages share each of four simplex channels. A fifth channel is dedicated to hospital-to-hospital communications only to ensure privacy.

The major types of medical traffic carried by the system are:

- routine medical consultations between the physician and the health aide;
- administrative calls to order drugs, supplies, etc.;
- emergency calls (an alarm at the hospital may be activated remotely by the village health aide);
- patient inquiries, i.e., follow-up on patients transferred to the hospital.

During the ATS-1 experiment, the impact of improved communication was dramatic. In the first year the number of days in which a health aide could contact a doctor increased more than five fold, and the number of patients treated with a doctor's advice more than tripled (Hudson and Parker, 1973).

The advent of the commercial satellite service with the dedicated medical network modeled on the ATS-1 system has extended these benefits to every village on a 24-hour per day basis. Impact on mortality and morbidity will take years to become evident, and is complicated by many other socioeconomic and environmental factors. However, the availability of emergency medical communication has helped to save lives, and the regular doctor call and administrative communication features have resulted in more village Alaskans being treated with a doctor's advice, a higher standard of performance by health aides, daily reinforcement and upgrading of their skills, and a more efficiently managed health care delivery system.

Total annual system cost for the medical network is $875,100 for 75 villages and 10 hospitals, or about $28 per site per day. However, the vast distances and difficult terrain make the cost of transportation very high in Alaska, so that the communication system becomes an attractive alternative as a substitute for transportation alone. Average airfare for a roundtrip from a village to a regional hospital is about $100. Consultations with the doctor by satellite, which make possible the treating of patients in the village rather than transferring them to the hospital, will save both transportation and hospitalization costs that amount to a minimum of $500 per patient, and may cost several thousand dollars. Transportation costs alone from remote islands exceed $600 (Hudson, 1981). Thus medical benefit-to-cost ratios range from at least 20 to 1 within a rural zone to more than 40 to 1 from the most isolated areas.

The Alaska Legislative Affairs Agency is responsible for providing information on legislative activity to the general public and for facilitating public participation in the legislative process, through legislative offices linked by a telecommunications network. Each regional office is equipped with audio conferencing equipment, a computer terminal for data transmission and electronic mail, and a telecopier to transmit letters, testimony, etc.

The audio conferencing network is intended primarily to allow local participation in legislative committee hearings by witnesses and interested observers in the communities served by the network, and also is used for informal meetings between constituent groups and legislators. The audio-conference capability is used daily, either for scheduled telemeetings or hearings, or for exchange of business among the various information offices.

The monthly cost to the State of Alaska for the entire audio network is about $16,000, or about $160 per hour, assuming 100 hours of use per month. State officials regard the system as very cost-effective because cost incurred with travel and per diem alone would be considerably greater if the same public participation were possible without the LTN.

Costs of testifying in Juneau in travel expenses and time away from home range from $500 to $1,500 for Alaskan citizens. Rural residents may now travel to their nearest local community to testify, at an average cost of $100. It is intended that in the near future they will be able to dial into the system from their own village telephone. Thus the ratio of benefits-to-costs in travel savings for rural residents can be from 5 to 1 to 15 to 1, in addition to the benefit of hearing from citizens who were previously unable to testify at all.

The state of Alaska is exploring a variety of applications of telecommunications as a means of addressing the educational needs of rural Alaskans. One service is an electronic mail system which now links the state Department of Education in Juneau with other state agencies in Anchorage and Fairbanks, five regional resources centers, and 52 rural and urban school districts. The system eliminates the need to send separate copies of messages to each site, and ensures that the ''mail'' will be received the same day. Mail sent through the postal service may take weeks to arrive because of infrequent transportation service or weather, and may often arrive too late to be of any use. Currently, the system handles about 100 messages per day, or about 2,000 per month.

The State of Alaska through the Department of Education and the University of Alaska is currently investigating other applications of telecommunications to extend educational opportunities to both school students and adults throughout the State. An educational teleconferencing network called LEARN/ALASKA has been established to enable groups of students and educators in different Alaska communities to participate simultaneously in university courses, staff development, professional meetings, etc. The LEARN/ALASKA network includes audio, freeze-frame video, videotext, and computer communications.

Public telephone service now reaches virtually all of Alaska's more than 200 permanent communities. Telephone traffic from the villages tends to be to the nearest regional center, while people in the regional center communicate primarily with major cities. The demand for toll telephone service in rural Alaska is higher than might be expected given the fact that most communities have only seasonal employment and/or marginal economic activities.

One of the major functions of the system is ordering of supplies for local retailers and for construction projects. In some cases, communication can be completely substituted for transportation, saving an average of $100 for transportation within a native region or from $250 to $500 or more for airfare and per diem if travel is required to a major city. Thus with a short intrastate call costing no more than $3.00, the ratio of benefits-to-costs for the rural telephone user can range from 30 to 1 to more than 80 to 1.

The telephone has also enabled increased participation in the political process and access to funding sources previously unavailable in remote regions.

Native organizations keep in touch with their rural constituents by telephone to inform them of new policies affecting them, to seek their direction, and to assist them in obtaining funding which they previously were unable to obtain because of the long turn-around in mail service. In addition to the savings in travel costs, this use of the telephone has increased the flow of development funds to rural areas and has given villagers greater control over policies affecting them.

Chapter 5

The Role of Telecommunications in Commerce and Industry in Developing Countries

Telecommunications can play an important role in entrepreneurial activities in developing countries. This chapter presents evidence on the effects of telecommunications in secondary and tertiary sectors in the third world, the relationship of telecommunications to transportation, and the employment generating impact of telecommunications, both in the sector and in the economy as a whole.

5.1   The Business Function of Telecommunications

Lesser and Osberg (1981) identify three basic functions of telecommunication:

- security functions,
- social functions, and
- business/economic functions.

They postulate that the business function has the greatest impact on socioeconomic development. They propose several hypotheses concerning the economic benefits of telecommunications:

1. Telecommunications, by increasing information flow, may improve the efficiency, i.e., productivity, of other factors of production.
2. Telecommunications, by increasing information flows, may improve the utility/effectiveness of various market outputs.
3. Telecommunications may expand the effective market area of business firms, giving rise to more efficient plant sizes and economies of scale.
4. Telecommunications may provide significant cost savings over alternative information-delivery systems.

5. Telecommunications, through the combination of the factors above, may enhance the socioeconomic development of isolated regions of nations or of nations as a whole vis-a-vis their external trading partners.

6. Telecommunications may permit significant reductions in inventory holdings through a centralization of inventories.

7. Government use of telecommunications may make possible significant improvement in the delivery of government services, in the profitability of government-run enterprise, and in efficiency of use of top-level personnel.

8. Government use of telecommunications may improve citizen access to government service.

9. Government use of telecommunications may facilitate contact and coordination between different levels of government in a country.

Lesser and Osberg compare three modes of information transfer: face-to-face, written and physically transported (which they refer to generically as "mail"), and electronic—or telecommunications.

From their analysis, Lesser and Osberg derive the following conclusions concerning telecommunications as a superior mode of information transfer:

1. In general, modal superiority will depend on many factors including type and urgency of message, efficiency of each mode, and characteristics of sender and receiver.

2. If we assume that each mode is operationally efficient within its own parameters and if we consider major types of messages, then the unique advantages of telecommunications appear in

   • information transfers involving long distances;
   • information transfers where the speed of message delivery (and response) is important;
   • information transfers involving "routine" information flows;
   • information transfers requiring limited participation of senior personnel;
   • information transfers involving persons who are unable or unwilling (for example, due to illiteracy or for reasons of social custom), to use mail (i.e., written communication), and where the cost of face-to-face contact is relatively high.

3. The advantages of telecommunications will be higher the more spatially dispersed the communicators.

4. In some situations, one or more of the operational characteristics (as opposed to cost characteristics) of a particular mode may uniquely qualify that mode for the message transaction. For example, a transaction requir-

ing a legal signature can only be carried out in writing (although this may be done face-to-face as well as by mail). For telecommunications, speed may be the single most important operational characteristic when distance and urgency eliminate other modes, i.e., it is physically impossible to send a written message or go in person *in the time available.*

5. In the more general case, there will be some trade-off between the modes. Elasticity of substitution between modes is defined as the sensitivity of the use-ratio of modes to changes in the relative price of modes. The ease of modal substitution in response to changes in price, (i.e., overall cost of use), will govern the marginal rate of technical substitution.

6. Attitudinal factors, traditional strategies of management, and/or ignorance of the capabilities of telecommunications may bias agents' modal choice, i.e., the marginal technical rate of substitution may be incorrectly perceived due to subjective factors.

7. Treatment of information transfer costs within organizations, and/or the failure to value time, personnel requirements, or support requirements may bias economic agents' perceptions of the relative costs of competing modes to the disadvantage of telecommunications, i.e., the correct "price" of alternative transfer modes may be incorrectly perceived.

8. Telecommunications may substitute for other information transfer modes; in a more dynamic framework, i.e., with the introduction of time, telecommunications may *also* complement other information-transfer modes.

Lesser and Osberg conclude that:

1. From a theoretical perspective, telecommunications confers benefits on society over and above the private benefits represented, or captured, by the price system.

2. In order for the socially desirable level of telecommunications investment to be provided, these public benefits of telecommunications must be taken into account.

3. If the externalities of telecommunications are not taken into account in determining the social profitability of telecommunications investment, telecommunications will be underproduced.

5.2 Impact on Secondary and Tertiary Sectors

The ITU GAS-5 manual stresses the importance of telecommunications to secondary manufacturing and to tertiary sector firms. A sector analysis of the importance of telecommunications indicated that 90% of telephone services in developing countries are used by subscribers in industry, business, banking, transportation, and government. For secondary manufacturing, telecommunications is needed to coordinate complex series of operations. The secondary sectors however, require telecommunications less than the tertiary (services) sector.

A study by the GAS-5 of the use of telecommunications in the U.S., Japan and Yugoslavia indicated that trade, finance, transport and service industries are the heaviest users of telecommunications. This importance derives from the fact that tertiary sectors cannot function on a day-to-day basis without telecommunications. Thus, it appears that telecommunications is necessary for these sectors to develop and expand.

Kaul (1978) reports on a pioneering study that examined the benefits of telecommunications through input–output analysis using an interindustry transaction table for 1968–69. Kaul found that intermediate consumption by industry amounted to 42.11% of total consumption of P & T services (with private consumption 42.93% and public or government consumption 14.96%). Of the intermediate consumption, almost 95% arose from the tertiary sector as broken down in Table 5.1.

Beal and Peppard, in a study sponsored by the Canadian Department of Communications, constructed a multidisciplinary model of telecommunications in a developing region, using northern Ontario as their example. The model included the interactive effect of tourism and telephony; an increase in tourism could lead to an increase in telephone usage which could help to stimulate an increase in tourism (Beal, 1976). However, another question could be asked: Would provision of a telecommunication infrastructure in such a developing region (in conjunction with other efforts such as advertising) help to stimulate tourism in the first place? Evidence from Alaska cited by Goldschmidt (1978) suggests that it would.

Studies by RITE (Research Institute for Telecommunications and Economics) of Japan (1978) suggest that telecommunications affect economic development primarily, but are necessary for social and political development as well. RITE divided 92 countries into six development stages, and concluded that the higher the development stage, the greater the effect of telecommunications investment. Therefore, they concluded that telecommunications was a more important strat-

Table 5.1
Intermediate Consumption of P & T Services
by Tertiary Sector, India 1968–1969[a]

Trade	67.75
Banking and Insurance	13.69
Education and Research	9.30
Transport	8.07
Others	1.19
	100.00%

[a]Source: Kaul, 1978

egy at the more advanced stages; however, the lack of telecommunications could be a bottleneck for economic development at any stage. This approach has some parallels with Hudson, who suggests an organizational take-off stage required for developmental media usage (Hudson, 1974). However, one wonders if RITE reached its conclusions because there are so few examples of telecommunications investment at the earlier stages.

The studies by researchers at the Moscow Electrotechnical Institute of Communication are concerned with the measurement of the microeconomic effects of telecommunications (see Gorelik and Efimova, 1977; Gorelik and Karaseva, 1975; Gorelik et al., 1978).[1]

Gorelik and Efimova (1977) state that long-distance communication enables matters to be decided on expeditiously and ensures a prompt reaction to users' information received. It saves time and helps to make production control more operational and flexible, to make the flow of production smoother, to speed up marketing, to reduce the volume of circulating capital and improve supplies, as a result of which production losses are cut, prime cost is brought down and labour productivity rises. All this eventually pushes up the country's national income.

The authors show how investment in the development of the trunk telephone system can result in savings in production costs. They conclude that 90% to 95% of the entire effect of use and development of the trunk system in the USSR accrues to the national economy beyond the confines of the communication sector.

Gorelik and Karaseva (1975) outline methodologies for studying the economic efficiency of the trunk system and estimation of results. Indicators used include saving of working time and personnel, increase in labour productivity, expenditure on production, and rise in output of the national economy. Surveys of several sectors showed that the effect of telephone use was related more to the size of the enterprise than to the sector. The authors conclude that the findings testify to the high economic efficiency of the long-distance system and may be used for traffic forecasting and capital investment decisions.

Gorelik, Efimova, and Karaseva (1978) propose further refinement of techniques to determine the economic efficiency of the trunk system. They use an estimation of the degree of interchangeability of the various types of communication (post, telegraph, telephone) and travel. Results from the study of 57 enterprises in 18 industries are cited. They calculated direct savings of time and indirect effects of telephone use resulting from greater management efficiency and flexibility. The authors show that the availability of trunk telephone service cuts the firms' expenditure on travel expenses by an average of 7 to 10%, and reduces the idle time of transportation equipment by an average of 10%. Another

---

[1] These studies in English translation can be found in "Soviet Studies on the Indirect Benefits of Investment in Telecommunications." (mimeo). OECD Development Centre, Paris, 1978 (Working Paper 5 of the Expert Meeting on Methodology for the Study of Socio-Economic Benefits of Telecommunications).

of the major benefits is the reduction by some 20% of the fines paid for late delivery of equipment and for substandard production. They calculate that each unit of 1000 calls results in a savings of 3,700 staff hours or the equivalent of 36,000 roubles in production costs. These savings are likely to increase as the quality of the telephone system improves. In 1980, the time saving per 1,000 telephone calls was anticipated to be of the order of 5,000 staff-hours, or 35% more than in 1975.

The income-generating effects of telecommunications are also significant. According to the Soviet researchers, total long distance telephone revenues amount to approximately 0.15% of gross social product, but the use of the trunk system generates 0.65% of the gross social product, and increases the productivity of labor by some 0.75%. The benefits of the telephone system are approximately 4.3 times higher than its costs, i.e., each ruble of capital invested in the long distance telephone system in the period of the tenth 5-year plan leads to a saving in expenditure of 4.4 rubles, as well as to an increase of 2.4 rubles in gross social product, or 1.2 rubles in national income.

Several studies carried out in Poland were, like the Russian studies, unknown to western researchers. A study in Poland by Hoffman (1975) came to the conclusion that in 1971, the indirect benefits of the telephone to Polish business users were approximately 13.5 times higher than the cost of using the telephone (i.e., annual rental plus communications charges). Hoffman also observes that if one takes into account the indirect benefits of telephone utilization, the theoretical period of return of capital investment in telecommunications is of the order of less than 1 year.

5.3   Telecommunications and the Business Firm in Kenya

A study by Tyler (1981) was designed to assess the impact of shortcomings in telecommunications services on the performance of selected businesses in Kenya and to quantify the benefits that would accrue to these business if telecommunications services were improved.

Shortage of investment capital has prevented the Kenya telecommunications system from expanding at a rate sufficient to satisfy demand from business and residential customers, a situation common in many developing countries. As a result, firms and other institutions typically suffer from both shortages of exhange lines and difficulties in placing and receiving calls at high traffic periods. The failure to achieve timely communications with suppliers, customers, and other parties contributes to personnel wastage, production losses, and other sources of inefficiency in the economic system, according to Tyler.

The Kenya study was a pilot project to develop and demonstrate reliable methods of measuring economic benefits of enhanced and expanded telecommunications services in a developing economy. It measured the extent to which nine firms would benefit from access to improved and extended telecommunications facilities. Each firm was surveyed to identify the various ways in which the effectiveness of the business was affected by the quality of telecommunications services. To provide the framework for the analysis, Tyler and his colleagues developed a series of models based on the premise that information is a necessary input to economic processes, and that communication services enhance the ability of organizations to obtain and transmit information. Thus information is considered to be a factor of production, like labor, materials, and energy. The models describe specific mechanisms by which the change from poor to good telecommunications results in an increase in efficiency. From these models a set of hypotheses was derived concerning mechanisms through which business performance could be affected by the quality or quantity of telecommunications services.

Table 5.2 is a summary of these mechanisms and the hypotheses of the impact of improved telecommunications. Nine different cost elements were identified for analysis:

1. The business expansion costs, representing the costs which the lack of effective telephone service imposes upon the firm by restricting its access to consumers, suppliers, and other producers. In practice, this expansion cost corresponds to the losses incurred by the firm as a result of its inability to profit from the economies of scale associated with greater production runs;

Table 5.2
Mechanisms Considered in the Study Through Which Quality and Availability of Telecommunications Services Affect Efficiency Results of the Firm[a]

Mechanism	Hypothesis on Impact of Telecommunications
Business expansion	Economic activities which yield a net economic gain (revenues exceed costs) can be operated on a larger scale and thus make a larger total economic contribution
Management time	Less time is wasted and hence management productivity is improved
Labor time	Productivity is improved through substitution of telecommunications for labor-intensive forms of communications
Inventory levels	Required levels of inventory are reduced
Production stoppages	Stoppages are reduced in frequency and duration
Vehicle fleet scheduling	Wasted mileage and vehicle-hours are reduced, fewer vehicles are needed for a given volume of work
Purchasing decisions	Buyers obtain better quality and price
Selling prices	Producers obtain better prices

[a]Source: Tyler 1981

2. The managerial time costs, which amount to the monetary value of the total amount of time lost by managers in making unsuccessful calls;

3. The labor time costs, which represent the cost of the extra labor used as a substitute for effective communication (e.g., messengers, drivers, etc.);

4. The inventory level costs, or cost of carrying additional inventories as a safeguard against the delays in reordering cause by poor telecommunications;

5. The production stoppage costs, which represent the additional time lost in production stoppages as a result of poor telecommunications. These production stoppages, resulting from machine breakdowns, shortages of spare parts, or unavailable raw materials, have a high marginal cost in the short run. They lead to substantial drops in revenue, but to very little offsetting savings in fixed costs;

6. The vehicle fleet scheduling costs resulting from the underutilization of vehicles and the higher mileage resulting from the difficulty of organizing back-loads (i.e., cargo for the return trip);

7. The purchasing decision costs, or costs resulting from the inability to contact a sufficiently large number of suppliers, in order to obtain the lowest price for raw materials, equipment, or services;

8. The selling price costs, which represent the reverse of the purchasing decision costs. The larger the number of potential customers which can be effectively contacted, the greater the possibility of selling the firm's product at a high price. (This factor appears particularly important in the case of exports of highly perishable agricultural products);

9. The supply costs which represent the costs incurred by the firm as a result of its inability, because of poor telecommunications, to offer on the market its total available supply of goods or services (such as vacant rooms in a tourist hotel).

Each of these nine cost items was computed in the enterprises concerned through a detailed analysis of telecommunications patterns, followed by in-depth interviews with the main telephone users in each of the firms.

The following firms were included in the study to represent manufacturing/processing, service, and rural enterprises:

- a large manufacturer of household and industrial supplies;
- a manufacturer of biscuits and other foods;
- a forwarder and stockist of machinery and industrial chemicals;
- a transport and forwarding company;
- a travel agent;
- a daily newspaper;
- a large vegetable exporter;
- a small horticultural business.

Tyler assessed the amount by which costs would be reduced and revenues increased if better telephone and telex services were available. The study concluded that the nine firms could save 1.2 to 9.2% of total revenues if they had access to improved communications. The average saving was 5% of gross revenues.

The magnitude of these potential benefits was compared with the cost of providing the needed extra telecommunications facilities. Using nationally averaged figures for the unit costs of providing telecommunications services in Kenya, Tyler found that gains from improving the telecommunications facilities available to each firm would exceed costs by factors ranging from 22.9 to 222.4. The average gain was over 100 times the corresponding cost of providing enhanced and expanded telecommunications services.

An analysis of the findings showed that the most significant consequence of improving telecommunications facilities would be to enable the firms to expand and take advantage of capacity utilization. Other important impacts would be lower production costs, less wastage of managerial and labor time, improved vehicle utilization, and more economical purchase and sales decisions.

Table 5.3 summarizes the ratios of benefits to costs of access to improved telecommunications services for the nine firms. This summary includes only those cost items which were relevant to, or significant for, the enterprise in question. Savings in production stoppages, for instance, are irrelevant to a hotel chain; and purchasing costs are unlikely to be very significant for a firm in the travel industry.

When examining the cost benefit ratios, it is important to realize that it would likely be impossible to improve substantially the telecommunications services for one firm without upgrading the network as a whole. Thus the cost of upgrading

Table 5.3
Summary of Cost-Benefit Ratios[a]

	A[b] Total Benefits	B Total Costs	A/B
Alliance	870,000	13,480	64.5
East Africa Industries	11,930,000	125,460	95.0
Industrial Distributors	2,390,000	17,080	140.0
Interfreight	5,565,000	57,900	96.0
Kenya Horticultural Exp.	2,627,500	31,460	83.5
Kenya Nurseries	27,400	2,560	10.7
House of Manji	4,417,900	14,760	299.3
Pan African Travel	275,400	2,320	118.7
The Standard Newspaper	6,035,600	31,060	194.3
Total	34,138,800	296,080	115.3

[a]Source: Tyler 1981
[b]Only those benefits greater than 0.05% of revenue are included.

service for a single firm assumes an investment in improving service for the national system. Tyler's cost estimates may therefore underestimate the true cost to the telecommunications utility.

The data show that the benefits of improved communications for businesses are likely to be substantial, but that the ratio of benefits to costs appears to vary widely from one type of enterprise to another. Also, the impact of improved telecommunications was less than might be expected in some cases such as stock levels which had been predicted to be lower with better telecommunications because of the capability of quickly ordering replenishments. However, in developing countries reordering supplies may not be equated with receiving them. For example, bottlenecks at the port which can delay goods for a month overshadow the time value of immediate contact with suppliers.

The study was designed as a pilot project to develop and test a method for improving knowledge of the linkages between telecommunications service provision and economic productivity. Tyler points out that the quantitative results are based on a very small sample and should be interpreted cautiously. Nevertheless, they provide strong evidence that in a typical developing country setting, the economic benefits resulting from the expansion of telecommunications services greatly exceeds the costs.

5.4   Communication and Transportation

The introduction of telecommunications services in a region can alter considerably its patterns of communication. Wellenius (1978) finds that three types of phenomena can take place:

- substitution for alternative means (such as transportation, mail, and telegraph);
- generation of new communication, which would not have developed without telecommunications;
- new requirements in other sectors (e.g., transportation) as a consequence of the increase in intensity and variety of interaction.

Wellenius cites the following evidence from a study in Chile:

> A survey of urban Chilean households in 1970 showed that, regarding communication among members of the family and with friends and kin, availability of the residential telephone tends to result in a partial substitution for the use of letters and telegrams by long distance telephone calls, a considerable number of telephone calls that constitute new communication events, and a net increase in visiting frequency. On average, families with telephones had more than three times as many of these communications per unit time as those without telephone. (Wellenius, 1978)

In Alaska, improved communications have made certain forms of travel, often hazardous and expensive in the Alaskan bush, unnecessary. For example, during the construction of a village high school, construction supervisors often had to fly out to call their suppliers. Such transportation is purely substitutable by communications. However, improved communication will not necessarily lead to decreased rural transportation demand. For example, the Aleutian village of Atka requested an airstrip for air service to replace its monthly mail barge following the installation of a telephone (Goldschmidt, 1978). Hudson (1974) also noted that improved communications in remote northern Canada also contributed to increased travel, as communications helped to facilitate greater economic activity.

It has been suggested that telecommunications development in rural areas may help counteract the rural exodus to the towns, a serious problem in many developing countries.

GAS-5 (1972 ed.) (International Telephone and Telegraph Consultative Committee 1972) notes that remote areas are handicapped by the high cost for shipment of finished products to the principal market areas, the high cost for shipment of tools and supplies manufactured in more favored regions, high costs due to traveling to and from locations of markets and supplies, delays measured in days when materials or information are required from the manufacturing centers, and lack of ready access to advice from a large pool of extertise. While the GAS-5 committee notes that most remote areas have primary commodity industries which use telecommunications only minimally due to the simplicity of their production, the use of modern telecommunications allows authorities to decentralize highly concentrated groups of industries and distribute them more evenly around the country: "The infrastructural development of telecommunications should stimulate the commercial and industrial life, thereby permitting efficient commercial and industrial companies to carry out their activities in places outside the traditional commercial centers existing today." GAS-5 cited as an example of such decentralization a Canadian firm which invested $400 million in a manufacturing plant in an isolated area with inadequate transportation facilities but which had an extensive microwave communications system serving the area. Thus, rural telecommunications development may allow a more balanced development of a country by allowing business and government agencies to operate outside of large urban areas.

In a study of the trade-offs between telecommunications and transportation, Ohlman (1981) tackles this issue of substitution using an information theory approach. Ohlman designs a requirements matrix which includes the following factors:

1. Fundamental system factors
    *   amount of information
    *   elapsed time

- distance
- funds available
2. Secondary system factors
   - type of information
   - reliability
   - storage
3. Human factors
   - availability
   - ease of use
   - dissemination/audience
   - interaction
   - language requirements
   - security

He provides five categories for each factor, so that the potential user can identify the required level of service for each category and compare the ratings of various media to determine the optimum choice.

Ohlman derives a "figure of merit" combining ratios of the four fundamental factors:

- message transmission speed, or distance per unit time (D/T)
- channel capacity, or quantity of information per unit time (I/T)
- message economics, or quantity of information per unit cost (I/C)
- distance economics, or distance per unit cost (D/C).

Ohlman's figure of merit then becomes $(D \times I)/(C \times T)$ (expressed in kilometers, bits, standard currency, and seconds.)

For standard size messages (300 words), the telephone has a figure of merit almost 200 times greater than the mail, and 20 times greater than telex. The shorter the message, the greater the advantage of the telephone. For very long messages, the mail gets the highest rating.

5.5   Energy and Telecommunications in Developing Countries

Most developing countries depend heavily on oil imports for their energy supply. There is widespread agreement that the oil importing developing countries (OIDCs) face an energy crisis which substantially harms their development programs. Generally, OIDCs are dependent on oil for a larger proportion of their commercial energy requirements than the developed countries—approximately 66%, as opposed to 50% for the industrialized market economies that make up the OECD.

Moreover, the changes taking place in the economies of the OIDCs as part of the development process—for example, the growth of urbanization, factory industry, and motorized transport—have meant that their growth in demand for commercial energy exceeds the rate of growth of their Gross National Product. Consequently, if present energy policies continue, oil imports will place increasing pressure on the balance of payments and indebtedness of the OIDCs. Without changes in energy policies, Tyler (1981) suggests that the present rate of growth in the OIDCs, which is still somewhat higher than the growth rate of the industrialized economies, will be undermined.

The predicted increases in expenditures on oil imports by the OIDCs are massive. Tyler reports that OIDCs spent US $50 billion on oil imports in 1980, representing approximately 3.3% of their GNP and 26% of their export earnings. Given the continuation of present growth rates in oil consumption by the OIDCs, this would rise to US $110 billion (1980 prices) by 1990, representing 4.5% of their predicted GNP, and 32% of predicted export earnings.

Tyler (1981) focuses on three main opportunities to use telecommunications to increase efficiency of energy use in OIDCs:

* to improve transport efficiency;
* to reduce travel needs;
* to improving the energy efficiency of industrial and agricultural processes.

Transportation typically accounts for 15 to 25% of direct energy use in industrialized countries, a similar or slightly higher proportion in middle income developing countries, and a little less in low income developing countries. Telecommunications technologies can have an important influence on the level and efficiency of energy usage. Tyler distinguishes between:

* the use of telecommunications as an adjunct to transportation systems, increasing their capacity of efficiency, and
* the use of telecommunications to substantially alter the level and composition of demand for transport.

The first of these possibilities applies largely to freight transportation. The second concerns reduction of the need and demand for travel.

Transportation accounts for 10 to 20% of total energy consumed in the low income developing countries, but according to Tyler this figure understates the importance of energy consumption in transportation. In most developing countries, road transportation accounts for the overwhelming share of total transportation, and it is totally dependent on imported oil. In addition, a large proportion of the freight transported consists of high bulk, low value agricultural and mineral products; up to 50% of the final price of these products consists of transportation costs. The potential benefits of telecommunications in increasing the efficiency of the transportation system are therefore very significant in developing economies. Tyler suggests that national energy conservation programs should give much greater weight to the potential contribution of telecommunications to the reduction of unnecessary freight vehicle mileage. They estimate that through better telecommunications, it should be possible for the oil-importing developing countries to save about $18 billion per year in oil import costs.

The analysis of the role of telecommunications in reducing travel needs is more complex. Telecommunications can indeed serve as a substitute for transportation; calling a person on the phone is much faster and cheaper than traveling 300 miles for a face-to-face meeting. This substitution or diversion function is very important, given the countless instances throughout the developing world of personal trips necessitated by poor or nonexistant telecommunications services. However, these trips are also used for a number of other purposes, for both business and personal matters. But if telecommunications can serve as a substitute for travel in specific, well defined circumstances (e.g., for routine communication with familiar colleagues, as opposed to complex negotiations with strangers), its use also tends to increase the total volume of communication, and could contribute to an effective increase in the total amount of travel, as noted by Wellenius (1978). Tyler calls this the generation effect of telecommunications. Both generation effects and diversion effects must be considered to measure the effective contribution of telecommunications to reductions in travel. From an earlier analysis of the experiences of NASA and a major chemical company in the United States, Tyler concludes that the generation effects, while relatively important, are nevertheless well below the gains accruing from substitution. The teleconferencing system developed by these two organizations led to a net reduction in travel of 25%, which he calculates resulted from a diversion effect of 35%, and a generation effect of 10%.

One of the most important observations by Tyler and his colleagues is that the potential of telecommunications as a substitute for travel is much larger in the developing countries than in the highly industrialized nations. In the latter countries, he postulates that the generally high quality and density of the telecommunications network has already led to a very significant substitution of telecommunications for travel, and the potential savings offered by yet more telecommunications are fairly limited. In the developing countries, the situation is very different. Introducing a telephone where there was none before can lead in the short-term to enormous reductions in unnecessary transportation.

5.6   The Impact of Telecommunications on Employment in Developing Countries

The telecommunications sector requires personnel to install, operate and maintain facilities and to manufacture telecommunications equipment. Blanc (1981) focuses on the direct employment effects of telecommunications, i.e., the total number of new jobs likely to be created between 1976 and 2000 both in telecommunications services (the operation and maintenance of the networks) and in the manufacturing of telecommunications equipment.

Blanc's projections of telephone demand and density for the 1976–2000 period cover 24 countries, representing approximately 42% of the projected world population in the year 2000. Two of these countries are centrally planned economies, 13 are developed market economies, and 9 are developing countries. Two basic hypotheses were considered. The first assumes that extensive use will be made of standard technologies, such as crossbar exchanges and conventional cables. The second hypothesis assumes that a much wider role will be given to the most sophisticated technology, notably electronic exchanges and optical fibers. These technological hypotheses represent technological extremes. In practice, most countries are likely to proceed with a mix of technologies.

Blanc concludes that the impact on employment of using advanced technologies is more substantial for the industrialized than the developing countries. This trend appears to result from the potential of these technologies to increase the growth rate of the telecommunications sector. For the four developing countries with projected growth rates in excess of 10% (India, Iran, Mexico, and Tunisia), total employment with highly sophisticated technologies will be higher than with labor-intensive technologies. This finding may suggest that the countries with the lowest telephone density should select the most modern telecommunications technologies. Choosing the sophisticated technologies which require fewer workers ultimately results in a higher growth rate for the whole network, thereby requiring more employees in the operation of the telecommunications system.

Blanc estimates the total number of new jobs likely to be created in the telecommunications manfacturing sector. He projects the total number of new lines to be installed worldwide between 1976 and 2000 at from 1.07 billion (low estimate), to 2.225 billion (high estimate). This corresponds to a yearly average of 42.8 to 90.0 million new lines per year. By the year 2000, the total number of new lines installed per year could be around 100 million worldwide.

These projections can be used to derive the number of workers in telecommunications manufacturing. Assuming that each person employed in equipment manufacturing produces around 20 lines per year, the number of workers needed to produce the 100 million lines a year projected for the year 2000 will amount to 5 million. This output per worker represents the average between low productivity telecommunications industries (10 lines per worker per year) and high productivity industries (30 lines per worker).

Another approach uses the value of production and the average output per worker. Assuming an average constant cost of $1,000 per line throughout the period, and an output per worker (measured in constant dollars) in the $15,000 range, the $100 billion of telephone equipment produced by the year 2000 (100 million lines per year at $1,000 each) would require around 6.66 million workers. With a higher productivity hypothesis ($30,000 per worker), the total number of workers would amount to 3.33 million. This higher productivity may be somewhat optimistic. A large part of the growth in telecommunications manufacturing facilities will take place in developing countries, which at present have only a very small telecommunications manufacturing industry with a rate of output per worker well below the $15,000 figure. It may therefore be more realistic to assume a worldwide average productivity figure in the $20,000 to $25,000 range by the year 2000. This would correspond to a total number of jobs in telecommunications manufacturing of 4 to 5 million, i.e., about the same estimate as that derived above from worker productivity figures.

This estimate of 4 to 5 million jobs in telecommunications manufacturing in the year 2000 can be compared to the total number of jobs in the industry in 1976 to obtain the number of new jobs likely to be created between 1976 and 2000. Estimates based on the number of workers in telecommunications manufacturing in the 20 largest multinational corporations and on indicative figures for the centrally planned economies suggest that total worldwide employment in telecommunications manufacturing in 1976 was in the range of 1 to 1.5 million. Using an average of 1.2 million, the total number of new jobs likely to be created in telecommunications manufacturing between 1976 and 2000 will be 2.8 to 3.8 million, or an average of 110,000 to 150,000 new jobs per year throughout the world.

It appears from Blanc's analysis that a large proportion of these jobs will be in developing countries. The implications of technology transfer are thus very significant. Already there is a shortage of trained staff at all levels of telecommunications administrations in many developing countries. It will be necessary to increase the number of training institutions and on-the-job training opportunities, and to ensure that training is included in contracts with suppliers either to procure equipment or to set up production facilities within developing countries. Failure to take these steps is likely to result in increase dependency on expatriate personnel.

Unfortunately, Blanc does not deal with the central question of the impact of telecommunications growth on employment throughout national economies. It would appear from analyses by Hardy (1981), Tyler (1981), and Berry (1981), among others, that technological change and growth in telecommunications will contribute to expansion of many economic activities in commercial sectors that could result in new jobs, but that there will also be major changes in the nature of many new and existing jobs, with increased emphasis on information production, distribution, and management.

Blanc (1981) also examines the relationship between telephone density and telephone utilization (measured in calls per telephone) in a sample of countries. He finds that countries can be divided into four groups:

- Group 1: countries with high telephone density and high utilization (primarily industrialized countries, and a few islands with high population density and dependence on service industries);

- Group 2: countries with rapid growth in telephone density and moderate utilization (primarily industrializing countries and oil exporting countries);
- Group 3: countries with low telephone density and high utilization, perhaps indicating high urban concentration of telephones and/or enclave economies;
- Group 4: countries where utilization appears independent of increase in density.

Given the hypotheses about the role of telecommunications in economic growth proposed by several authors reported in this book, it would be interesting to classify a large sample of nations using these indicators, and then to examine trends in economic growth. It could be hypothesized that economies of developing countries in Groups 2 and 3 would be expanding at a greater rate than countries in Group 4.

Jéquier (1983) makes preliminary estimates of the impact of telecommunications investment on national employment using data from the study by Hudson, Hardy, and Parker (1981). He converts the economic surplus generated by telecommunications investment into job equivalents. Jéquier uses the results of the study by Hudson et al. (1981) which estimates the increase in GDP resulting from installation of telephones or thin route satellite earth stations in regions with different telephone densities. Using a ratio of average earnings per capita to GNP per capita, Jéquier estimates the number of job equivalents that would be generated by the predicted increase in GNP.

In the poorest region with a per capita income of $100, he calculates that each telephone installed would generate 12 jobs. However, the impact of telecommunications investment on GNP is highest in low income countries, and decreases as the level of income increases. Also, the indirect effects on employment are not linearly related to increases in GNP; as per capita GNP increases, average wage levels and thus the cost of new jobs also increase. In Group 3 countries, with a per capita GDP of $245, the installation of each new telephone would result in one job over 9 years, whereas in Group 2 countries (with a per capita GDP of dollars) each telephone would generate 0.25 jobs, or it would require four telephone installations to generate the equivalent of one job.

Jéquier points out that the cost of job creation as a result of telecommunications investment is relatively low, if current figures for the cost of thin-route satellite services are used. Investment in telecommunications may help to achieve the goal of low cost job creation as well as other social and economic development goals.

While it appears that impact of telecommunications investment on GNP and indirectly on employment increases as telephone density decreases, Jéquier also suggests that there is a critical mass of telecommunications investment which will stimulate additional economic activity. He suggests that once a country's telecommunications density reaches 30 to 50 lines per 100 population, i.e., so that virtually every firm and organization and most families are connected to the network, new employment opportunities arise. For example, many service sector industries such as banking, tourism, and airlines require extensive communications facilities, as do new services including teleconferencing, home electronic services and shopping, and videotext.

Jéquier concludes that there are likely three job generating effects of telecommunications. The first, as analyzed by Blanc above, is direct job creation for production of telecommunications equipment and installation, operation, and maintenance of the network. The second, as used in Jéquier's estimates, results from the overall increase in economic activity made possible by telecommunications investment. The third is the job generating effect of new telecommunications-intensive services characteristic of a highly developed industrial economy.

Part III

Issues in Research and Planning

Chapter 6

Methodologies for Further Research

The purpose of this chapter is to outline various methodologies that may be applied to research on the role of telecommunications in socioeconomic development. Particular attention is paid to the collection and analysis of data from field settings. Problems and techniques of aggregate data analysis are also discussed.

6.1   The Value and Limitations of Case Studies

It is probable that future research on telecommunications and development will profit more from microstudies than from macrostudies of telecommunications and national development. The microstudies offer the advantage of allowing fairly precise development of theory and method within a readily testable environment. Macrostudies, while more ambitious in scope and useful when offering strong results, are difficult to perform at this time largely because it is difficult to formulate theory specifically given the general lack of information necessary for developing theory, difficult to formulate rigorous methodologies, largely due to the lack of theory, and difficult to acquire accurate data of national, regional, and local indicators outside of a handful of highly industrialized nations. It is possible that a series of microstudies of the quantitative impacts of telecommunications on specific economic sectors will provide the basis for developing larger models which will prove useful in the future for planning telecommunications investments.

It is difficult to determine to what extent the case studies and sector studies cited in the literature review can be generalized. There are at least as many differences as similarities between developing regions. Then differences, including economy, political structure, culture, and history, may profoundly influence the impact of any innovation, including communication. On the other hand, the common problems of isolation, lack of infrastructure, lack of adequately-trained personnel, and needs to stimulate and diversify rural economies may indicate that techniques developed in one rural area may be applied to advantage in another.

A second problem of case studies is the difficulty of controlling for all of the variables that may affect the outcome. A new industry may stimulate migration; a shift in government policy may create different economic incentives; a bottleneck in the transportation system may limit the effectiveness of logistical communications.

A third problem is the constraints which may limit or influence the type and extent of indirect benefits from telecommunications. The nature of the economic activity, the political organization, the social organization, and the variety and values of the cultures may all affect how telecommunications systems are used and with what effect. The problem is, therefore, not only in controlling for these variables as well as for other variables, but also in determining to what extent findings from a case study can be generalized.

6.2   Some Problems with Indicators

Several of the studies of investment in telecommunications in various countries use the Gross National Product (GNP) or Gross Domestic Product (GDP) as the key development indicator. However, it is important also to recognize the limitations of the GNP/GDP statistic as a development indicator.

One problem is that aggregate wealth cannot adequately measure quality of life. Although it can be assumed that in most cases higher per capita wealth/ productivity should correlate with the quality of life indicators, such is not always the case. With increasing concern among development agencies about ''basic human needs'' and about improving conditions of the poor majority, new indicators are required that give a more accurate reflection of quality of life.

The U.S. Overseas Development Council has proposed a simple ''Physical Quality of Life Index'' (PQLI) which uses three basic indicators: life expectancy, infant mortality, and literacy, rated on a scale of 1 to 100 and consolidated into a single index. Ratings of countries using this index show that poorer nations may have higher ratings than some resource-rich nations. The Overseas Development Council states:

> These divergences from what the expected relationship might be indicate that significant improvements in basic quality of life levels can be attained before there is any great rise in GNP; conversely, a rapid rise in GNP is not in itself a guarantee of good levels of literacy, life expectancy, or infant mortality. (*Economic Impact*, quoted by Pierce, personal communication, 1978)

If telecommunications is to be examined within the framework of a development theory which considers social as well as economic factors, then an index such as the PQLI should be included in statistical analyses. However, the quality of life index has the same fundamental problems as other development indicators—it does not show a distribution throughout the society:

> The indicators are only proxies, they rarely measure directly the equality of the service or social components being measured, and they *cannot* show the access of the general population to them. There is no international standardized nomenclature in the social development objectives . . . .

> Neither single indicators nor composite development indicators are applicable as general policy directives because there is no supporting framework of static or dynamic theoretical analysis. (Michael Ward quoted by Cruise O'Brien, personal communication, 1978)

We must therefore treat any national statistics with caution, as they will not indicate conditions among the disadvantaged. If the focus of the planning or analysis is to be on rural development, however, a step closer can be taken if data can be obtained for the region or rural province/state under consideration.

The same statistical problems apply to telephone data as well. The national telephone density (or number of main stations per hundred inhabitants) gives no indication of distribution of telephones throughout the population. It is likely that the phones are clustered in urban areas with rural areas having little or no service. Within urban areas, telephones are likely to be available only to higher income residents. Therefore, more detailed statistics are required on distribution of telephones within a country. Statistics on telephone density in urban areas only, for example, will indicate whether private telephones are available only to the upper income strata and institutions/businesses. For public telephones, the community or neighborhood may be a more appropriate unit of analysis. For example, statistics or how many villages of "x" population or more have at least one telephone will give a much better profile of telephone availability.

Even the proximity of the instrument, however, does not guarantee functional access: if there is a breakdown anywhere in the system, the circuits are overloaded, or if the transmission quality is excessively poor, the telephone might as well not exist. Therefore, data on quality of service in various subregions are required to give a meaningful picture of telephone distribution. (See Section 6.3 on Access.)

The utilization factor (UF) is an index based upon the correlation between GDP and the number of telephone stations, expressed as number of telephone stations per $100,000 of GDP. However, as Okundi (1978) points out, the UF can hide real growth in telephone density if there is instability in GDP. For example, in Kenya, telephone density was .66 per 100 inhabitants and UF was 6 in 1970, while in 1977 the density had increased to .95 while the UF had fallen to 4.2 due to growth in GDP stimulated by high world prices for agricultural products such as coffee and tea.

6.3   An Approach to Resolving Indicator Problems: An Index of Access

In order to overcome the problems with traditional telephone indicators, an index of access is proposed. The components of this index include (based on Hudson, 1974; Hudson and Parker, 1975):

Physical Distance: The actual average distance customers must travel to use the telephone . . . (e.g., from the home to a store or community office to where the public telephone is located). In order to minimize this distance, telephone utilities in industrialized countries have sought high penetration of households as well as businesses. However, in developing regions, the relevant unit may be the community rather than the household or office. For example, in Alaska and northern Canada, communication planners have used the isolated community as the unit of analysis. The State of Alaska's goal is to provide at least one telephone to each permanent community of 25 inhabitants or more. Similarly, Prasada (1976) proposed that increased attention be given to *community telephones*. In urban areas, subdistricts may also be chosen as the planning unit so that there is at least one telephone within a specified walking distance of each resident. With the higher density of urban populations, consideration must also be given to the population density, so that additional criteria based on population may also be required.

Cost: The average cost to the user of making the most common types of calls; this cost should also be expressed as a function of average disposable income in the community or region. Note that several types of data are required to get a meaningful indicator of cost to the user. It is not enough to simply know the price of a local call. The price must be related to the income of the users. Telephone pricing statistics that compare the cost of a local call from one country to another have little meaning unless they take into account the income of various user groups within each country. An Ethiopian may pay only 8¢ per call, while an American pays 25¢, but that 8¢ represents a much greater proportion of the Ethiopian's disposable income. In rural areas, the local call may not be the appropriate unit for consideration, if most rural calls are long distance, requiring some toll charge or flat rate in excess of a local urban call. Here some basic traffic data will be needed to show the major calling patterns or communities of interest which should be taken as the basis of the cost calculation (e.g., the price of a three minute call to the nearest market town or administrative center).

Quality of Service: A telephone system is only as reliable as its weakest link.

If for any reason the telephone instrument does not function, it might as well not exist.

The leader of an isolated Cree Indian village in remote northwestern Ontario, Canada, pointed out that although a telephone was installed in his community, his people effectively had no access to it because it did not work:

> Why build an expensive microwave system of towers to bring telephone service into an isolated community like Angling Lake, and then fail to maintain the single pay phone that is the end result of the whole project? (quoted in Wa-Wa-Ta, 1978)

Telephone utilities keep statistics on quality of service generally compiled in annual surveys, including:

- blockage rates on toll trunks;
- mean time to get dial tone;
- outage hours per trunk per year;
- mean time between trouble report and repair completion.

It is again important to note that it must be possible to break such statistics down by subregions to get a meaningful indication of access to service. For example, the aggregate statistics on quality of service kept for Ontario, the most populous and industrialized province in Canada, do not reveal quality of service problems in the small isolated northern communities.

Sociocultural Factors: A working telephone in a village or urban neighborhood still may not be accessible if the local people are unaware of its existence or apprehensive about using it. Factors which may enter into this sociocultural distance include:

- *Location of the telephone:* a phone in a local store or post office may be accessible to virtually the whole population because everyone is familiar with these locations. However, a telephone located in a police station or government office may be inaccessible to people who are apprehensive about entering such "official" places.
- *Awareness and skill:* The public must be aware that the telephone exists, where it is located, and how and when it may be used. The public must have the basic skills necessary to use the equipment (e.g., a radio telephone will not be accessible to the public if they are not aware of how to operate it; a modern dial phone will not be accessible to people who do not know how to find a number and dial it).
- *Assistance:* Any of the above problems can be overcome if the customers can be assisted by operators/local agents who speak their own language and are willing to explain telephone procedures. The most modern dial

pay phone may be inaccessible if the only source of information is a telephone directory that the customer cannot read and an operator who cannot speak the customer's language or will not take the time to explain procedures.

6.4   Hypotheses and What They Suggest About Research Methodology

The studies outlined in this book suggest the following hypotheses about the role of telecommunications in socioeconomic development:

- The effects of telecommunications use do not accrue exclusively to the users, but accrue also to the economy and the society in general.
- Telecommunications permit improved cost-effectiveness of rural social service delivery.
- Telecommunications permit improved cost benefits for rural economic activities.
- Rural telecommunications permits more equitable distribution of economic benefits.
- A certain level of organizational development and complementary infrastructure is required for socioeconomic benefits of telecommunications to be realized.
- Telecommunications use can facilitate social change and improve quality of life.

Lesser (1978) provides a framework for analysis through a discussion of certain characteristics of communications:

1. *Externalities:* ''Externalities are benefits or costs which are captured by, or imposed on, a party other than the immediate producers or consumers of a good or service.''

Lesser notes the following problems raised by the externality characteristics of telecommunications:

- extension of telephone services into an underprivileged area may benefit the developed region more by allowing it to exploit the markets of the underdeveloped region;
- because of significant external benefits an estimation of demand may not be taken as a proxy for benefits;
- it is difficult to estimate the benefits of investment of telecommunication in a region as investment in one region may benefit another and, ''by the

same token, the total benefits derived from telecommunications in one region may not come entirely from the plant investment in that region.''

2. *Heterogeneity:* Telecommunication differs in types of users, types of use, and types of system hardware:

Because telecommunications is not a homogeneous good, in measuring the benefits of telecommunications there is a need to differentiate the benefits gained by different types of users, to differentiate the benefits generated by the type of information flow being sent, and to incorporate an understanding of the system technology into these assessments. (Lesser, 1978)

3. *Complementarity:* Lesser points out that telecommunications may play an important but accommodating role in economic development rather than a causal role. But if telecommunications can act as a cause of development, much more careful attention must be given by planners, and much more detailed understanding of this role is required. However, telecommunications does not likely act in isolation:

This notion of complementarity between telecommunications and other infrastructure components may in turn mean that the economic development benefits of telecommunications are to be mainly realized only in combination with other infrastructure components. A corollary hypothesis might be that there is some minimum threshold level of each such infrastructure element which must be realized before the benefits of any one element can be fully maximized and that to expand one element, say transportation, beyond its threshold while leaving some other, say telecommunications, below its threshold, will produce a minimal, or zero, impact on growth. (Lesser, 1978)

These hypotheses and characteristics of communications suggest the following about design of field research:

1. Research focus must not be limited to individual telephone users.
2. To capture social benefits, designs must include institutions that could be expected to benefit from telecommunications use: health delivery systems, educational institutions, extension agencies, etc.
3. Key economic activities in the region and the community must be identified. Research must include individuals and organizations involved in these activities, e.g., trading, small farming, fishing, local businesses and industries.
4. Purposes, frequency, and communities of interest of telephone use must be identified for various classes of users.
5. The telephone services available, and access criteria including their location, availability, and pricing structure must be identified.

6. Telephone facilities available and information on their reliability includ-
    ing blockage rates, outages, signal quality, etc., must be documented.

6.5   Research Designs

Three basic research situations should be considered for field studies:

1. *Rural regions where telecommunications services are to be installed for
    the first time*

From a research point of view, this is the ideal situation because it allows
measurements to be made of key indicators before and after the installation of
telecommunications facilities, and allows for monitoring of trends over time.
Research designs can be chosen to allow comparison with comparable communities and/or regions without telecommunications.

2. *Rural regions where telecommunications facilities have been installed to
    support social and/or economic activities*

Some regions may appear particularly fruitful for study because it appears that
the telecommunications systems installed are being used for developmental purposes and/or that data collection will be simplified or enriched by the availability
of accessible historical records of telephone traffic and usage. Here, before and
after analysis will not be possible, but it may be possible to collect historical data
for several time periods, to compare with similar regions without telephone
service, and to make at least anecdotal comparisons with conditions before
installation of the service.

3. *Case studies*

Certain sectoral applications of communications may appear particularly
worth studying, especially if data can be collected on several applications in the
same sector.

Examples from studies reviewed in this book include agricultural applications
in the Cook Islands, business applications in Kenya, educational applications in
the South Pacific, and health care applications in Alaska. The problem with case
studies, of course, is their replicability. To the extent that a good research design
can be used and measures may be standardized for use in other settings, the
results will be more likely to be generalizable, at least to the sector or type of
application.

Although it is not possible to impose a fully rigorous experimental design in a
field setting, quasiexperimental designs have been developed that control most

threats to reliability and validity. Designs which could be imposed for field research in telecommunications include:

1. *Pretest-posttest designs*

In studying regions where telecommunications service is to be introduced, it will be possible to collect data to measure variables that are hypothesized to be influenced by access to telecommunications both before and after the introduction of the service. Early studies have suffered from lack of a design to control for other factors, such as introduction of other services or investments, shifts in the economy, population, etc., all of which might account for changes in the measured variables. Therefore, it is important to control for as many of these factors as possible by introducing control sites. If the plans for installation of service are already fixed, there may be no choice in the experimental communities except to draw a representative cross-section from the population of sites to be served. The control group can then be chosen from communities that are (if possible) similar in several significant dimensions such as population, economic acitivites, cultural/linguistic composition, and degree of isolation. It is important to note that communities that will receive telecommunications first are likely to differ from others in one or more of these dimensions. However, simply having a control group which can be pre- and posttested will provide more confidence in determining whether changes were due to the introduction of telecommunications. This process produces a nonequivalent control group design which controls for most threats to reliability and validity.

The design can be strengthened if sites can be randomly assigned into experimental and control groups. This may occasionally be possible if plans for installation are not fixed and there is no political or economic penalty associated with the order of installation (for example, with a satellite system where it is not necessary to extend wires or build towers along transportation routes.) This process of randomization would yield a fully experimental pretest–posttest control group design that could control for every major threat to internal validity.

2. *After-only designs*

Where the telecommunication system has already been installed, after-only designs must be used. One cannot simply measure certain variables in locations with and without telecommunications and have any confidence that differences are attributable to use of telecommunications. An after-only design can be used confidently if the groups are randomized. However, this will of course not be possible where facilities are already in place. One way to introduce some control into this design is to do a panel study—i.e., to measure both groups at several points in time. This approach could be used for studying the effect of telecommunications in an industry (e.g., fishing) or in a health care system, where it is possible to observe comparable systems or parts of one system or industry with and without communications at several intervals. This method would therefore be appropriate for strengthening sector case studies of telecommunications effects, e.g., in health, education, agriculture, fishing, etc.

6.6   Measurement of Key Variables

To obtain satisfactory evidence concerning the external benefits of rural telecommunications using the research design proposed above, three kinds of variables must be measured for each community at each key time period—i.e., the communication variables, the organizational variables, and the economic variables.

  1. *Communication variables*

  Physical availability and reliability of communication services. For each community it will be useful to know the number of public long distance telephone circuits available; reliability of toll circuits and local equipment; and the availability of alternate communication channels, including postal service, telegraph, mass media channels, and transportation facilities that could be used to convey messages. The changes in these variables are a presumed cause of the projected economic and social benefits.

  Utilization of communication services. Traffic data including volume of calls, traffic patterns (community of interest data), and peak calling periods will be important to collect. Unused channels are unlikely to bring social benefits. Changes in level of utilization (which may be dramatic if the prior lack of facilities suppressed demand) presumably lead to whatever benefits are obtained. Channel utilization measures also reveal congestion factors where access is restricted and future expansion of use is inhibited by busy circuits or other network blockages.

  The purposes of use. The purposes for using a system are likely to have a significant impact on the kinds of benefits. If most of the usage is for social calls, the results can be expected to be different than for business calls. Automatic recording of traffic statistics may be possible, but content data are likely to depend on questionnaires, interviews, or log data. Where there is a single toll station, a simple log may be kept by the attendant to record the date, time, destination, and general purpose of calls (see Hudson, 1974). Interviews with callers during a sample period may also be used. It may be important to interview a sample of *all* households in the community to ascertain differences between users and nonusers. Gathering data by recording or listening in on a sample of calls may be tempting, but constitutes an unethical and unwarranted invasion of privacy.

2. *Organizational variables*

Since the research hypothesis predicts that benefits from telecommunications investment will occur when appropriate social organization is concurrently available, it will be important to obtain descriptions of the kinds of social organization present in different locations at different time periods. The size and location of various institutions and organizations (schools, health clinics, cooperatives, businesses, government offices, etc.) may be affected by the availability of communications services. They may also constitute a complementary variable, necessary for benefits of telecommunications to occur. Changes in these organizations or institutions over time and especially the emergence of new organizations should be noted.

In addition to such "institutional" measures, another kind of organization should be examined: the communication patterns measured by evidence concerning who talks to whom with what frequency. At the aggregate data level, geographically dispersed "communities of interest" can be measured by aggregate measures of the volume of calls connecting different communities. At the individual level, it may be useful to include interview or log items asking who was called (or communicated with by other means) with what frequency. Changes over time in the network of who talks to whom may provide significant clues to changing social organization. To be useful it may be necessary to distinguish calls or contacts by purpose of call (at least distinguishing business from social networks).

3. *Economic variables*

Evidence of local or regional economic activity is the essential criterion variable necessary to demonstrate (or fail to demonstrate) external economic benefits of telecommunication investment. Aggregated national economic statistics are unsatisfactory for the purpose; the unit of analysis for the economic variables must be at the same level as the communication and organizational variables, that is, the local rural community or district. Various measures may be attempted: agricultural output, tax collections, income surveys (asking income questions in interviews), employment surveys or records, credit and banking activity, etc. Various kinds of data are likely to be available at different levels of aggregation in different countries. Careful exploration of possibilities will be needed to uncover data sources and to arrange for their availability. If all else fails, economic questions asked of community residents and representatives of institutions can provide the basic data, i.e., new data collection can substitute for analysis of records when the records don't exist or are unavailable.

To conduct the cost-effectiveness trade-off analyses concerning alternate ways of delivering social services to rural areas, budgets of organizations will be needed for analysis. Unit costs for different levels and kinds of services are required in order to calculate budgets required under different circumstances (e.g., with telephone supervision permitting less highly trained and paid staff to perform certain activities, or with communication substituted for transportation in some service areas). Examination of actual shifts in budget allocations after the introduction of telecommunications would permit observations about the extent to which theoretical benefits were in fact realized.

A variety of statistical analysis procedures is likely to prove useful in summarizing, reporting, and drawing causal inferences from field study data. Correlations, partial correlations, comparative time series, analysis of variance, and other statistical procedures may be necessary to extract sufficient evidence to justify the necessarily expensive data collection. Different analyses would be required for different policy purposes and audiences. Some of the data could be immediately useful as feedback to the telecommunications project implementers. Additional analyses may be useful to national policy planners considering possible extensions to other parts of the same country. Yet other analyses and descriptive reports may be necessary to provide evidence by which planners in other countries could make judgments on whether comparable results could be expected in their countries.

In examining the role of telecommunications in development, most researchers use various means of measuring and comparing benefits and costs. Kaul (1981), for example, uses a consumer surplus approach, which has also been used in studies carried out by the World Bank. Kaul's research focuses on the benefits to the individual by determining to what extent the benefits of telephone use exceed the costs of alternative means of communicating, and on defining the consumer surplus as the difference between the cost of telecommunications and the cost of physically delivering the message. Tyler (1981) and Kamal (1981) both develop benefit/cost ratios which include a variety of direct benefits and substitution effects of using telecommunications. Tyler (1981), Kaul (1981), and Hudson (1981) examine transportation costs in terms of personnel time and fares for commercial and service activities, and compare them with the costs of telecommunications use. All of these approaches have considerable validity, but more standardization of these measures would help to make comparisons across projects and across sectors.

However, all of these approaches miss at least some of the intangible or ephemeral benefits of telecommunications. For example, the value of a life saved as a result of communications in an emergency will not be measured solely by the economic contribution of that person to society. And the sense of well-being and security resulting from a telephone call from an isolated worker to relatives at home will not be captured by the cost of the call or even a decision to remain at the isolated post longer than planned.

6.7 Evaluation of Telecommunication Projects

Such detailed research will not be feasible on most telecommunications projects in developing regions. However, it should be possible to collect evaluative data which can indicate to planners the extent to which the system is meeting the users' needs and contributing to socioeconomic development.

Gellerman (1978) reports the evaluation requirements for rural telecommunications projects funded by the Inter-American Development Bank:

1. *Rural public telephones*

- Monthly summaries of number of main telephones in service, total volume of use and total revenue.
- Annual survey for a 2-week period of sample locations showing name of community, population served, distribution of population by time or distance from telephone, distance to next nearest public telephone, distance to nearest exchange, waiting time, purpose of calls, destination of calls, time or distance from caller's home to telephone, duration and cost of calls sent and received.

2. *Rural exchanges*

- Monthly summaries of number of exchanges in service, number of main telephones in service, volume of usage, revenue broken down into monthly service charge, local, long distance, and installation fees.
- Annual survey for a 2-week period of sample exchanges showing local and long distance usage broken down by origin and destination, as well as data listed above for public telephones.
- A one-time survey of subscribers indicating number in family, occupation by head of household, years of schooling, and income.

These data are to be collected during the execution of the project and for five years after completion.

This level of data collection should be possible for most telecommunications projects, drawing upon research skills in the country, for example, in the government or in universities. The basic telecommunications traffic data should be collected by the telecommunications administration.

Where feasible, efforts should be made to impose some form of control to isolate benefits, using the quasiexperimental designs outlined above. For example, interviews should be conducted in communities with and without telephones, or within the telephone communities, with a sample of residents that would include users or subscribers and nonusers or nonsubscribers. In this way patterns of use and perceived benefits can be identified. Particular attention should be paid to institutional users or subscribers—e.g., local leaders, representatives of local businesses, field staff of government agencies in the district, local social service workers—as it has been hypothesized that their usage may result in collect benefits.

Efforts should be made by researchers and funding agencies to develop standard coding categories—e.g., for user profiles, calling frequencies, and purpose of calling. Results could then be compared across projects and across countries.

6.8   Aggregate Data Analysis

Another tool for examining the relationship between telecommunications and development is aggregate data analysis. In this book, the studies by Hardy (1980) and by Blanc (1981) and studies using Hardy's methodology (Parker, 1981; Hudson et al., 1981) all used aggregate data analysis.

The unit of analysis for these studies was the nation. Data for aggregate variables (e.g., telephones per 100 population, GDP per capita) for a number of different nations at one point in time is called a cross-section. The analysis of the data base is *cross-sectional analysis*. The data on variables for one nation for a number of time periods are called a *time-series*. *Time-series analysis* is the term used for working with such data. Data for aggregate variables for a number of different nations at multiple points in time are called a *cross-sectional time-series*.

Given aggregate data to work with, whether cross-sectional, time-series, or both, there are a number of ways to analyze the data.

The major means of analyzing these data fall into a class of methods based upon the general linear model, including correlations, partial correlations, and multiple regression. It should be noted that the general linear model encompasses a wide variety of techniques which allow a large number of possible research strategies to be employed. The methods examined here are those most often used, or those which are logical extensions to prior research.

At the most basic level, the data can be used for descriptive purposes, e.g., to compare nations according to indicators of development in various sectors including telecommunications, and to compare growth trends in different nations.

Bivariate relationships among variables can also be examined. For example, GDP per capita and telephone density can be plotted for several nations for a single year. A bestfitting line can be obtained to describe the plot. The slope of the line indicates how many units of GDP per capita change when telephone density changes by one unit. It is tempting to use such an equation to predict how increases in telephone density might affect economic development of a nation. However, this would require assumptions that might be unwarranted, i.e., that there was a causal relationship between the two variables, and that the process which relates telephone density to economic development is equivalent for all nations, both industrialized and developing. Similar unwarranted assumptions would be required to interpret the relationship of telephone density to economic development in a single nation using plots of these variables over a multiyear period.

Because of their limitations, such analyses should be regarded as exploratory and descriptive. No causal inferences can be made based on such correlational evidence alone. The directionality of the relationship cannot be ascertained, nor can one rule out the possibility that evidence relationships are spurious, being caused by some other variable or variables.

Such challenges suggest the application of techniques which consider multiple independent variables and/or time lags between independent and dependent variables.

Partial correlation is a technique which allows work with more than one independent variable. This technique allows us to test for spurious relationships between two variables. In partial correlation, the relationship between two variables is calculated while holding any other variables of interest constant. This approach helps to control for spuriousness, but does not address the question of causality.

By using time-series data it is possible to obtain empirical evidence of the dominant direction of the relationship between two variables. Given data on two variables (e.g., GDP per capita and telepone density) for a nation over a number of time periods, it is possible to compute lagged correlations between the two variables.

Lag time or *lag interval* refers to the number of time periods by which the values of one variable follow those of another variable. For example, if our analysis uses telephones in 1950 and development in 1955, the lag time for development is 5 years. *Lead time* or the *lead interval* is the number of time periods by which a variable precedes another in an analysis. In the above example, telephones would have a lead time of 5 years.

The general procedure would be as follows. We first compute a correlation for data on GNP and telephones at the same time period. We can then compute a correlation for data for GNP at a time period before that of telephones, two time periods before telephones, three time periods before telephones and so on. Correlations can also be computed one time period after telephones, two time periods, three time periods, etc.

If a significant correlation is obtained for GNP leading telephones, this indicates that GNP is causally prior to telephones. If a significant correlation is obtained with telephones leading GNP, we have evidence that telephones are causally prior to development. We may also find evidence that both variables are causally prior to each other. One reason for this double causal priority would be that the variables are mutually causal.

Discovering causal priority between two variables, as well as a correlation still does not assure us that one variable is the cause of the other. The correlation found may still be spurious. We can again test for spuriousness by statistically holding other variables of interest constant.

Cross-lagged correlations can combine the lagged variable approach with tests of spuriousness. The simplest use of cross-lagged correlation involves cross-sections of two variables at two points in time. These are called *cross-lagged panel correlations*. For example, with data for many nations on GDP per capita and telephones per 100 for 1960 and 1970, the generally accepted procedure would be to correlate development in 1960 with telephones in 1970, and telephones in 1960 with development in 1970, using the nations as the cases. The higher of the two correlations would indicate which variable is causing the other. To test for spuriousness, it would be possible to use partial correlations to control for other variables that are possible causes of observed relationships.

Both *simple* and *multiple linear regression* techniques can be used for making predictions of the value of a single dependent variable, from the value(s) of a single or multiple independent variable(s).

With cross-sectional data, values could be used for GDP per capita, telephones per 100, radios per 100, newspaper circulation per capita, and mail volume per capita over a number of different nations at one point in time, for example. This approach is similar to Cruise O'Brien (1977). Multiple regression analysis can be used to develop an equation combining these media variables into a predicted value of development. Statistics are available to estimate the predictive accuracy of such an equation. Similarly, an infrastructure index might be constructed including indicators of telecommunications, transportation, water, electricity, and/or social services to test the infrastructural complementarity hypotheses outlined above.

With time series data, values could be obtained for these same variables for one nation over multiple time points. Multiple regression could be used to create an equation relating those media variables to development. Predictions of the level of development for that nation based on the media variables could then be made.

We must be careful in making causal inferences from both the time-series and cross-sectional analyses. Although highly accurate predictions may be obtainable, the causal influence of the independent variables cannot be assumed, unless all plausible alternate explanations have been refuted.

For example, we must be cautious in interpreting the results of the studies by Berry (1981) and Blanc (1981), which show interesting correlations and projections using aggregate variables, but do not control for all other factors which might explain or influence the relationships.

Hardy (1980) elaborates on the discussion above, and provides an explanation of the use of structural equation models or path analysis for making causal inferences from aggregate data.

A number of aggregate data sources can provide useful information for examining telecommunication's role in economic development. Telecommunications statistics are available from a number of sources.

AT&T's *The World's Telephones* is published annually by its Long Lines division. The series goes back to the early 1900s. Data provided in each publication includes total telephones, telephones per 100 population, and the number of international and domestic calls made for every nation of the world. Total telephones and telephones per 100 population are also broken down for major cities in each nation. However, there appear to be inconsistencies in reporting from one country to another, and there is no verification of the data submitted.

ITU's *Yearbook of Common Carrier Statistics and Radiocommunication Statistics* has yearly coverage. Variables reported here include population; number of households; telephone, telegram, and telex traffic; number of telephones; and revenues and investments from these telecommunications services.

A number of sources for aggregate data on social and economic variables exist. The World Bank's *World Tables* includes economic variables such as GDP per capita, GDP by economic sector, income distribution, and government expenditure by economic sector. Social indicators such as literacy, calories per capita, physicians per capita, birth rate, and population are also to be found. Coverage of these indicators varies from reports by year to five- and ten-year intervals.

The *UN Statistical Yearbook* and the *UNESCO Statistical Yearbook* contain demographic variables such as population, size of nation, and population density. They also have data on mass communications media, but not on telecommunications. The *UN Statistical Yearbook* does contain data on economic variables, but the range of variables is not as large as that from the World Bank. The *UN Statistical Handbook* also does contain data on the number of telegrams sent within a nation.

The *World Handbook of Political and Social Indicators* and the *Cross-National Time-Series* (also known as the *Cross-Polity Survey*) are two of the most extensively used data sets in aggregate political and social analysis. The *World Handbook* lists values for variables at five-year intervals beginning in 1950. The *Cross-Polity Survey* has continuous coverage starting no later than 1960 for its variables. Both sources have similar variables such as GNP per capita, percent of population in different economic sectors, literacy, life expectancy, and urbanization. Variables concerning mass media and telecommunications (telephones and telegraphs) also exist in these sources.

Although much useful information can be found in these sources, they have severe limitations for answering the questions raised throughout this book concerning the impact of telecommunications on *rural* development. To make inferences concerning the role of telecommunications in rural development, we require data which at minimum give indications of the economic conditions of the population living in rural areas and the telecommunication services available to them.

At the simplest level, we would need a measure such as telephones per 100 population in the rural areas, and per capita income for the rural population. Then, using the techniques outlined above, it would be possible to begin to ascertain strengths and directions of relationships between the variables. Other variables would be required to test for the spuriousness of any discovered relationships. The lack of urban/rural distinctions in much aggregate data from developing regions hinders the progress of research in using aggregate data analysis techniques to examine the role of telecommunications in rural development.

We are therefore forced to return to case studies for much of the research on telecommunications in rural development, and to search for rural regions about which good economic and telecommunications data are available. With existing data and research techniques we can obtain good estimates of the degree to which telecommunications investment in a nation influences its overall development. However, we cannot get precise information on the extent to which telecommunications alone or in combination with other elements can contribute to development.

Chapter 7

Planning Telecommunications for Rural Development

7.1   Demand and Need for Rural Telecommunications

As has been noted earlier, rural telecommunications are likely to cost more per capita and generate less revenue than comparable investments in urban or interurban services. There are not likely to be strong financial incentives for investing in rural telecommunications. One approach to increasing the attractiveness of rural telecommunications investments would be to reduce costs. Some strategies for minimizing costs and optimizing system design for rural services are outlined in the following sections. Another approach would be to take a broader view of need than is indicated through simple demand projections by taking into consideration the indirect economic benefits of telecommunications investment.

It is commonly assumed that demand for rural services can be estimated using standard forecasting techniques. Yet, these projections tend to underestimate actual demand for rural services, as demonstrated by the volume of calling once services are installed. For example, in northern Canada and Alaska, rural and isolated residents pay three to four times as much annually as their urban counterparts for long distance calls, despite the fact that these people have significantly lower average incomes than urban residents, and many of them function in a semisubsistence economy. But they recognize the value of being able to communicate with relatives and friends, to order supplies, compare prices for their fish and furs, contact government agencies, etc.

Despite the fact that demand in terms of actual telephone usage may exceed projections, it is obvious that rural and sparsely populated areas will not generate the volume of traffic that would come from more populated areas. Yet the need for communications may be significantly greater, because rural residents have few alternatives to using telecommunications; they must rely on mail delivery that is often slow and/or dependent on the weather, or spend a substantial amount of time and/or money to travel to convey or locate the information in person.

Is that need significant enough to justify the expenditure on telecommunications services? Clearly, this is a complex question that involves an analysis of the cost of providing service, opportunities to share costs with other users, and realistic forecasts of traffic growth. But telecommunications supply and demand should not be seen as forming a closed system where revenues must necessarily cover costs for each separate component of the system. The benefits derived from the system may greatly outweigh the costs. The problem is that many of these benefits are not effectively measured, nor are they assigned any weight in the determination of system economic viability. They do not show up on the balance sheet of the telephone company or telecommunications agency. Yet there must be recognition at the national level that telecommunications is important to national development and the achievement of national goals and policies if the necessary services are to be provided.

7.2   Demand Forecasting for Developing Regions

It is not the purpose of this book to review in depth telecommunications planning techniques. However, the chapters of functions and benefits indicate that in planning telecommunication infrastructure for developing regions, several factors should be considered.

Gimpelson (1976) states that "underestimation is endemic to voice communications forecasting; this applies to international, national, and local service growth." It appears that, particularly in rural and remote areas, telecommunications authorities tend to underestimate demand for telephone services. Parker (1978a) cites the example of Alaskan village residents queued to use the single community pay telephone. In the Canadian Arctic, Bell Canada has encountered blockage rates of over 20% in Inuit communities where local exchanges have been provided with long distance service by satellite (Inuit Tapirisat of Canada, 1978). And in northern Ontario, long distance toll revenues jumped by 300% to 800% when reliable satellite or microwave service was installed in remote Cree Indian communities (Wa-Wa-Ta, 1978).

Part of the problem has been lack of data on which projections can be based. Setting goals based on GDP may be unrealistic. Waiting lists may also be unrealistic indicators of demand as many potential customers may be too discouraged by long delays and high installation costs to bother to apply (Wellenius, 1978).

In an earlier work, Wellenius (1971) criticized the typical techniques of demand forecasting in developing countries (using waiting lists as the indicator or expressed demand or attempting to compare needs and integration level and population). He then proposed a technique for collecting and sampling illustrative cases and extrapolating the results to all centers in the classification.

Gellerman and Ling (1976) find that energy sector forecasts can be used to forecast telephone demand in some cases:

At least on a priori grounds there exists a threshold value of energy consumption beyond which an electrical subscriber has a potential to become also a telephone subscriber.

This approach may well be useful in some countries where residential electrification has preceded telecommunications. However, it will not help forecast demand for public telephones serving a village or neighborhood. Also, there may be instances where telecommunications investment precedes electrification. Such has been the case in some parts of northern Canada where telephone service has been provided to log cabins where families use kerosene lanterns and wood stoves. In this instance, telephone penetration may be an indicator of demand for electricity.

7.3   Optimizing Telecommunications Systems Design for Rural Services

Because of the greater distances to be covered and lower population densities, rural telecommunication systems generally cost more than urban systems to install and maintain. A study conducted for the ITU/OECD research project by a team from CIT-Alcatel (Jeancharles et al., 1981) calculated that on the average the total cost of a rural telephone subscriber line is five times higher than an urban line. The greatest cost is attributed to connecting equipment, i.e., the link between the individual subscriber and the exchange. Smaller exchanges are more expensive per line because of more limited economies of scale, and rural buildings may be more expensive if materials must be transported to rural locations to construct them.

Higher costs are generally coupled with lower revenues per line in rural areas than in urban areas, often resulting in a gap between capital and revenues which must be closed by subsidizing rural services with urban and interurban revenues (internal cross-subsidy) or by direct subsidy from the government. It is not a generally popular policy to set rural rates to recover rural costs.

There are, however, several steps which can be taken to reduce costs. Small solid state digital exchanges or modified PABXs have been developed for rural use that are highly reliable and require very little power. Radio links (VHF or UHF) may be less expensive means of linking rural subscribers to exchanges than stringing wire or cable. And use of part of a locally available building such as a school, community hall, or hospital to house telecommunications equipment or construction of buildings using local materials may all help to reduce capital costs.

Jeancharles et al. conclude that there is no single best technical solution, and that the design of rural networks calls for a mixture of technologies selected as a result of an analysis of several parameters including terrain, population density, distances, power supplies, and network configuration.

Some pioneering efforts to develop planning tools which do include such factors as geography and demography, investment capital available, other communication service requirements (such as radio and television, dedicated narrowband systems), etc., have been made by Stanford University's Communication Satellite Planning Center. The Stanford group has emphasized the creation of tools that could help developing country planners determine whether communication satellites should be considered, especially for thin route telephony for rural areas. Particular attention has been paid to the need to integrate satellite facilities with existing terrestrial facilities (see Sharma, 1976; Pinheiro, 1976). Two other reports explore optimized utilization of the satellite for thin route services—through development of demand assigned multiple access (DAMA) system for thin route telephony and planning (Sites, 1976) and optimum utilization of preassigned and DAMA services (Velasquez, 1976). These planning tools use computer modeling to provide speed, accuracy, flexibility, and generality. The models employed are quite complex, in order to reflect actual conditions as faithfully as possible. The four major programs are a satellite system program, a small station location determination program, a large station location determination program, and an orbit-spectrum utilization program. The first program takes account of the demand for circuits over a specified period and determines the least-cost configuration of the space segment and ground segment. The second program determines the least-cost allocation of small earth stations for a region which has virtually no terrestrial interconnections. It provides for interconnection of all communities by the most economical method, ranging from open-wire cable to HF radio, and direct placement of a small earth station. The third program determines the least-cost method of accounting for growth in demand for capacity in areas which are now presently served by terrestrial facilities, or for which service is planned for with terrestrial facilities in the near future. The result is an installation plan for major earth stations at critical locations nationwide. The fourth program determines minimum-cost satellite system from the standpoint of providing a wide range of services under the constraint of maximizing the use of the orbit-spectrum resource. The orbit spectrum is a nondepletable, constantly renewed resource, and the strategies for maximizing its use are also described.

7.4 Satellites as Appropriate Technology for Rural Telecommunications

The experiences with rural telecommunications outlined in previous chapters indicate external social benefits beyond the internal economic return on rural telecommunications investment. However, the investment required for rural terrestrial facilities may be beyond the reach of many developing countries.

For many rural areas, especially those in more remote locations or separated by more rugged terrain, satellite communication may be significantly more economical than terrestrial alternatives. In most cases, the least-cost solution will involve a hybrid system combining satellite and terrestrial technology (e.g., rural radio telephone). It may be difficult to think of communication satellites as appropriate technology for rural telephony, especially since most satellites have been designed and used as substitutes for undersea cable or terrestrial microwave on heavy traffic international and interurban routes. Even though costs per circuit for the satellite (space segment) and the size and costs of ground stations have been declining rapidly in recent years, most operating experience has been with systems that are inappropriately expensive for thin-route rural communications. Nevertheless, the general characteristics of communications satellites make them ideal for rural thin-route applications when technical specifications appropriate to those applications are selected.

The most obvious characteristic is the cost-insensitivity to distance. The cost of reaching the most remote community is virtually the same as reaching nearby communities. A major corollary of that characteristic is that communication capability can be installed in order of priority of need, independent of location. Instead of installing service first in communities nearest existing facilities (even though they may need it least), more distant locations, which have greater needs and higher costs of travel as a substitute, can be given service first. This complete flexibility of location may be important in coordinating communication installation with the specific needs of rural development projects.

Satellite systems are likely to be more reliable, more robust, and easier to maintain than terrestrial systems, an important consideration in rural areas, especially in mountainous or jungle or desert terrain. The space segments of satellite systems have proven themselves to be highly reliable, as they must be in orbital locations inaccessible to repair services. The ground segments, because they are located near the human settlements they are intended to serve, are more accessible for maintenance and repair than are the remote repeater sites necessary for much terrestrial communication. Because they can reach any other point in the network in a single "hop" through the satellite, reliable interconnection depends solely on the two stations involved. This contrasts with terrestrial systems which depend on a series of sequential links, the failure of any one of which can disrupt the connection. This characteristic makes the satellite system as a whole very "robust," because a failure in any ground station affects only that location and has no negative impact on other parts of the satellite system.

Satellites also permit a degree of flexibility with respect to capacity that is impossible in terrestrial systems. Ground stations can be installed to provide as little as a single voice channel of service using single channel per carrier (SCPC) equipment. Additional capacity can be added easily in those locations where it is required, as demand develops. Other services, including data transmission and radio and television reception, also can be added incrementally to the same basic ground station as demand requires. They can be installed in exactly those locations where the demand develops without having to be installed elsewhere. This contrasts with a terrestrial microwave system that must be designed at the outset at all locations to provide for the maximum end-to-end capacity required. It also contrasts with terrestrial open-wire systems which cannot have major capacity expansions at any remote location without major cost throughout the entire system.

Satellite systems also permit a simplex circuit (half of a standard duplex telephone circuit) to be shared at a number of locations as a common conference circuit. This capacity is less costly than duplex circuits interconnecting each pair of points. Unlike terrestrial conference circuits which are technically difficult because of different power levels on different lines, this capability is easier to provide by satellite than standard telephone service.

If thin-route Single Channel Per Carrier (SCPC) earth stations are equipped with Demand Assignment Multiple Access (DAMA) capability, then satellite capacity does not need to be dedicated to each rural earth station. When needed, an unused circuit can be selected from a pool of general purpose circuits. By the use of simplex conference circuits and thin-route DAMA equipment, the space segment costs for rural communication can be made quite small relative to the ground station costs and relative to all-terrestrial alternatives.

7.5   Satellite Systems Optimized for Rural Services

The communications satellite offers a particularly promising means of reducing rural transmission and interconnection costs. However, as we have seen, at present satellite systems tend to be optimized for communication between cities, i.e., using relatively large and expensive earth stations carrying large volumes of traffic. Even the new generation of higher power satellites is primarily designed for television distribution (e.g., direct broadcasting) rather than for interactive thin route (low density) telephone service.

The ITU has proposed a satellite system concept designed specifically for this thin-route rural application. The proposed system, known as GLODOM for global/domestic system, would consist of four in-orbit satellites placed to cover the entire globe, and particularly all the developing countries. The system would be optimized for this thin route rural domestic service, but could be used for interurban and regional traffic as well as for radio and television distribution (Pierce and Jéquier, 1983).

Earth stations proposed by Nickelson and Tustison (1981) would consist of a 3-m. antenna and mount, power system, radiofrequency and baseband electronics, and demand assignment multiple access (DAMA) system to improve the efficiency of transponder utilization. The components are chosen to minimize power consumption so that the earth station can be run off a storage battery, which in turn could be powered by solar cells (photovoltaics). Nickelson and Tustison also propose a new exchange design with high capability logic and a low power consumption which increases incrementally with activity (i.e., with negligible idle power consumption).

The proposed system would have 48,000 telephone channels (4 satellites x 24 transponders x 500 telephone channels) and 48,000 earth stations. Each earth station would be connected to an average of 20 telephone lines, so that the system would consist of 960,000 telephone lines.

The ITU estimates that the space segment cost will amount to $300 million, and that each earth station will cost about $20,000 manufactured in quantity, so that the entire system will cost approximately $1.3 billion. Assuming a total capacity of almost 1 million lines, the cost per line would be approximately $1,300, which Pierce and Jéquier point out is not much different from the average cost per line in urban areas. Therefore, they conclude that it is possible to bring telephone service to rural areas at a cost not significantly different from the cost of serving urban areas.

Pierce and Jéquier further point out that this estimated capital cost is comparable to the cost of a steel mill or automobile factory, and about half the cost of a large fertilizer plant, oil refinery, or gas liquefication plant. The estimated capital cost of the GLODOM system represents only 4.5% of one year's concessional aid granted to the developing countries by the industrialized countries belonging to the Development Assistance Committee of the OECD (which granted $28 billion in concessional aid in 1979).

The purpose of the GLODOM system is to provide universal access to telecommunications. The basic criterion is that no one should be more than 5 km. or one hour's walk away from a telephone (this criterion was adopted for rural telecommunications by the government of India; see Kaul, 1981).

Pierce and Jéquier estimate that revenues from the system could actually cover amortized space segment and earth segment costs because the cost of a 3-minute call, assuming a fully operational system, would be from 10¢ to 30¢. Of course, higher rates would likely be set to cover operating and interconnection expenses.

It should be remembered that the benefits of such a system are likely to exceed greatly the actual revenues generated by its use. Hudson et al. (1981) projected indirect benefits in the form of increase of GDP of $367,000 per satellite earth station over ten years in regions of very low telephone density (.01 or 1 telephone per 10,000 inhabitants). See Chapter 3.

It may be that a satellite system dedicated exclusively for rural use may not be the worldwide solution to rural telecommunications needs. But the GLODOM concept could be applied to domestic and regional satellite systems which are optimized for thin-route service to provide service for rural residents at much lower costs when integrated into an optimally designed rural network than would use of terrestrial technologies alone. Some parts of the world such as the Pacific islands may best be served by sharing a satellite with more major users such as East Asian countries or Australia, while Africa might support its own GLODOM type system.

7.6   Technology Transfer

Chasia (1976) points out that technology transfer is a social process which involves adjustments of a society's institutions to the technology. He distinguishes three levels in the transfer process:

- ability to use the technology;
- ability to operate and maintain the technology;
- ability to invent and make the technology.

Communications planners must weigh their options to determine what level they should attain. Clearly, any technology transfer must include the first level if it is to be functional and the second level if the country or region is not to be completely dependent on outside expertise. The third level might better be broken into several sublevels, including assembly of components, manufacture of components, and systems design.

Telecommunications planners must consider these technology transfer factors as well as demand forecasts in planning systems for developing regions.

Simpson (1978) states: "It takes more than technology to make viable telecommunications services available to society. In addition to technology, it takes appropriate economic, institutional, and social arrangements."

The first requirement for the implementation of the GLODOM system or any other rural telecommunications project would be training of the personnel in developing countries who would install, operate, and maintain the system. Many of these tasks would not require advanced education, and much of the work could be done by rural residents themselves, with appropriate training.

There is also a need for more trained telecommunications planners who can design systems for the needs of their own countries, and can adapt, modify, or redesign technologies to suit their own economic, geographical, and cultural environment. Special training courses, such as could be developed by the ITU for in-country use, may be the best solution. Opportunities to learn about new technologies and planning tools from industrialized countries and to share experiences with colleagues from the developing world may also help the diffusion of relevant information and stimulate innovation to serve developing country needs.

7.7   Constraints Limiting Benefits of
         Telecommunications

Clarke and Laufenberg (1981) identify major constraints which should be taken into consideration in projecting the benefits of investing in rural telecommunications. Macroconstraints apply across all sectors and appear to be sensitive to movements in the economic cycle, structural conditions which are slow to change, and shifts in relative balances between productive sectors. Among these macroconstraints are:

- regional dispersion of population, which may mitigate against rapid and low-cost extension of telecommunications and access to all rural producer groups;
- regional dispersion of economic activity, which requires a nationwide integrated communication system;
- patterns of income and asset distribution, which may require special subsidization policies to reach poorer rural residents;
- supply shortages and weaknesses of existing infrastructure, such as roads and power supply, which may influence choice of telecommunications technology and limit the benefits derived;
- telecommunications investment policies which may not correspond to national policies weighted toward rural development;
- foreign exchange availability, which may be the dominant factor in determining telecommunications investment policy.

Clarke and Laufenberg also consider a number of sectoral constraints which may restrain rural telecommunications development in Africa. They include:

- inadequacy of staff and of training facilities;
- scale effects in investments which cannot always be effectively optimized over time, so that demand may grow linearly, but capacity growth follows a step-like function, leading to problems of underinvestment or excess capacity;
- difficulties in demand estimation, particularly in rural areas without any experience with telephone services;
- need for organizational and institutional adaptation to rapid growth in telecommunications, which may require reorganization of telecommunications administrations;
- lack of capacity for adequate planning for telecommunications growth;
- financial constraints including unwillingness to subsidize rural services and inappropriate rural tariffs;
- inappropriate financial practices such as uncollected revenues which limit resources available for investment.

Clearly, these factors must be recognized in planning and evaluating telecommunications systems for rural development. Some of the sectoral constraints may eventually be overcome through more and better training in planning and management.

7.8   Telecommunications Planning for Development

This book represents an attempt to synthesize current knowledge on the role of telecommunications in development and to outline directions for further research. Telecommunications planners should be able to use the results of this research to help determine how telecommunications may contribute to national development goals. However, more attention must also be paid to developing techniques to assist planners in designing telecommunications systems which promote development needs. Factors to be taken into consideration include the purposes for which the system will be used, population density and distribution, geography, existing facilities, technologies available, and cultural/ organizational factors. Guidance may be needed in estimation of volume and type of demand and selection and for the transfer of appropriate technology for developing regions including local plant, toll interconnections, and power supplies.

The planning tools and techniques outlined in this chapter are efforts to meet these needs. However, more effort is required to develop and refine tools for telecommunication planners in developing regions.

If there is any validity to the evidence and hypotheses put forward in this book, the conclusion must be drawn that telecommunication planners and funding agencies cannot act in isolation. Telecommunication can be a tool to help achieve national economic and social goals. However, for maximum benefit to be derived, the telecommunication system must be integrated into national and regional development plans. Therefore, telecommunication planning should be included in planning of other development sectors such as infrastructure (transportation, electrification), social services (health, education, etc.), and agricultural and industrial development. Similarly, funding agencies must not consider telecommunications in isolation, but as one of many strategies for development, and as a key support to other sectors.

Coordination is required in development planning at the funding agency level and at the national and regional level among development planners and potential users of the system if maximum developmental benefits are to result from telecommunications investment.

Chapter 8

Conclusion

8.1 Recurring Themes

The studies summarized in this book add to our knowledge of the role of tele-communications in rural development, and bring out several recurring themes including the following:

1. Access

Simply having a telephone in a community does not guarantee that it is accessible to potential users. Kaul (1981) found that telephones in post offices were locked up after work hours when many villagers wanted to make calls. Hudson (1981) found that two-way radios in government offices in the Canadian north were not considered accessible by native residents, and that many communities in Canada and Alaska faced a major dilemma in finding the most accessible location for a village phone. (Apparently installing a phone outside in India is not considered a viable option because collections are manual rather than via a coin box; in northern Canada and Alaska, the harsh climate eliminates the option of installing the pay telephone outside.)

Access, however, is a two-way street, or should be. Public call offices (PCOs) may be an attractive solution to providing at least minimal telephone service in rural areas. However, this solution seems to assume that most telecommunication traffic will be from the rural areas to the regional or urban centers. It is extremely difficult to communicate with a village if there is only one phone for a village, or for several villages. In northern Canada, native people have found this system unworkable because it requires messengers to run around the village, or necessitates several calls to get the parties together. They have demanded local exchanges with telephones in residences and offices to overcome this problem.

In countries where this level of service is not an option, people often tend to use telegrams which are sent to the village and then delivered in person. Neither Kaul nor Kamal considers rural telegram service. In Egypt, telegrams are dictated over the telephone to villages without telegraph offices (this service is called a phonogram). This approach does convey the message, but is very time consuming and also ties up the trunk lines.

2. Rural Demand

Demand for telephone service in rural areas is almost always greater than would be predicted using standard estimating techniques. In rural India, Kaul (1981) found villagers willing to spend a significant portion of their income on telephone calls. In northern Canada, native people spend 3 to 4 times as much on long distance calls as the average Canadian, despite their much lower incomes. Similar patterns are found in Alaska, with village residents spending more than three times as much on toll calls as residents of Anchorage. Although rural telecommunications may not often be profitable, revenues generated are almost always much greater than planners anticipate, so that investment in rural services may not be as financially unattractive as expected.

3. Benefits are Related to Distance and Density

Kamal (1981) shows that benefits of telephone use are greatest for provinces most distant from major cities and for villages most distant from regional centres. Similar results have been found by Hudson (1981) for northern Canada and for a medical network in Guyana. Demand in remote areas appears to be highly inelastic, i.e., unrelated to price of a call, because of the lack of alternatives. The greater the distance from communities of interest, the greater the savings in travel costs and time in using the telephone. The implication for telecommunications planners is that a higher priority should be placed on providing services for remote regions.

Kamal points out that benefits are also related to telephone availability. More rural exchange lines and PCOs result in greater benefits. Thus there may be good reason to invest in a rural telephone infrastructure beyond single PCOs per village if appropriate technologies can be used so that benefits continue to exceed costs of providing service. (Cairo University is developing low-cost technology to extend rural telephone service.)

4. Communities of Interest

Destinations of calls vary with village economic activity and social structure. However, it appears that most rural residents call primarily within their own region. Kaul (1981) found that 63% of calls terminated within the district. Goldschmidt (1978) found similar results for traffic between Alaskan villages and regional centers. Hudson (1974) found that the major communities of interest in northern Canada were regional administrative and commercial centers, and that village traffic also clustered according to extended family links. Kamal (1981) found in Egypt that the education level is related to administrative level of communication; the more highly educated call beyond the district to provincial capitals or major cities. An understanding of rural communities of interest can be very useful to telecommunications planners in designing systems with capacity appropriately configured to meet projected demand.

5. Quality of Service

Service quality remains a major problem in rural areas. As Clarke and Laufenberg (1981) point out, expectations of low rural revenues lead to relatively less concern for rural quality of service, which in turn means that rural residents do not rely on the telephone and use it less than might be expected, creating a self-fulfilling prophecy. Infrequent maintenance and repair as well as high blockage rates seem to be problems confronting rural telephone systems worldwide. Poor quality of service means that benefits of rural telecommunications may be overestimated in regions where quality of service is poor; on the other hand, potential benefits may be underestimated if adequate service is available.

8.2   Unresolved Issues

Along with the recurring themes are unresolved issues, not always made explicit, that we must address to understand the role of telecommunications in the development process more fully.

1. Take-off Points

Both Kaul (1981) and Clarke and Laufenberg (1981) postulate that rural telecommunications has greater impact at certain stages of economic development. Kaul suggests that telecommunications becomes important when rural modernization begins, while Clarke and Laufenberg find a step function which indicates that telecommunications has a greater impact at certain stages of economic development. Other researchers suggest that communications play a complementary role in development along with other inputs and infrastructures. Further research on these issues could provide better guidance to planners on when and where to concentrate their rural telecommunications investments.

2. Who Really Benefits?

Although research has added considerably to our knowledge of benefits or rural telecommunications, the question of who really benefits remains unclear.

Does the fact that traders and proprietors benefit from telecommunications lead to increased benefits for the village as a whole? Does the ability of a village wholesaler to contact retailers in the city benefit the wholesaler, who may adjust his profit margins according to demand, or the peasant farmer who may have few alternative outlets for his crop? It appears that the social structure and economic organization of rural communities needs to be taken into consideration in evaluating the benefits derived.

3.   The Need for Consistency and Replication in Research

A variety of approaches to a problem can often be valuable in discovering new insights and relationships. However, they also lead to different results. For example, there are several important variations between Kaul's (1981) and Kamal's (1981) studies. If Kaul had interviewed only key officials, and used the scheme for analyzing benefits developed by Kamal, he would have arrived at different results. Conversely, if Kamal had used data from a wider sample of telephone users or chosen a different approach from the counterfactual to estimate benefits, he too would have arrived at different results. It will be important to replicate these and other approaches in other rural settings.

4.   The Effect of Institutional Constraints

Several references have been made to the value of these results for telecommunications planning. However, Clarke and Laufenberg (1981) point out that there are many constraints which limit the ability of telecommunications administrations to plan and provide services adequately for rural areas. One problem may simply be institutional barriers which make it difficult to plan for national needs as a whole, or to develop tariff and service policies that actually reflect national goals.

The countries studied here may provide useful examples. In Egypt, the major priority has been to improve Cairo's telephone system. Only recently has there been a commitment to improve rural telephone services, which are in many locations so congested that rural residents do not try to use them. In India, Kaul (1981) estimates requirements for approximately 40,000 PCOs to meet India's basic rural telephone needs. It has been demonstrated that in many cases satellite systems may be the least cost solution to providing such services. In fact, India now has its own domestic satellite. However, this satellite has been designed primarily to carry heavy route traffic between major cities and to serve a few extremely isolated outposts. It will also have a transponder to provide rural television services similar to those demonstrated using the ATS-6 satellite. With a different approach to the satellite design, the same earth station that receives television in a village could also have been used for telephone service.

These examples are not meant to be overly pessimistic. They simply point out that the question of applying telecommunications for rural development goes beyond an analysis of relative costs and benefits.

8.3   Social Values of Rural Telecommunications: The Intangible Benefits—A Concluding Note

Research which focusses on the economic value of telecommunications tends to miss or underrepresent the social benefits of telecommunications, whether for emergency assistance or contact with relatives and friends. Although perhaps methodologically intangible, these benefits are nevertheless quite real. And these social functions may have developmental benefits as well. For example, in remote northern Ontario, Indian trappers rent two-way radios to take out on the trapline. They use the radios to communicate from their camps to their home villages in order to keep in touch with their families and to request help in emergencies. In the evening they chat from camp to camp. The radios have proved their worth in human terms many times when injured trappers have been evacuated following a radio call. But there are other benefits. Trappers now are willing to stay out on the trapline longer because they can stay in touch with their communities. And they are willing to take their families along, knowing that the women and children can also get help in emergencies.

Similarly, evidence in many parts of the world indicates that telecommunications boosts morale of service workers posted in rural areas, such as teachers and nurses. There is evidence that availability of links to family and friends as well as for emergency assistance can contribute to lowering the rates of staff turnover in remote posts.

Finally, as has been pointed out in studies by Hudson (1981) and Goldschmidt (1978), among others, telecommunications services can contribute to improved quality of rural life. Staying in touch with friends and family, being able to order supplies from the city, and knowing that expert advice is available in emergencies can help overcome the disadvantages of isolation.

Perhaps the best summary of the intangible value of rural telecommunications comes from a doctor posted on a remote island in the Cook Islands who was interviewed over HF radio by the author:

Researcher: "What did you do before you had the radio to get medical advice? Over."

Remote doctor: "Just prayed to God. Over."

References*

Abler, R. 1977. The Telephone and the Evolution of the American Metropolitan System. *In* I. de S. Pool (Ed.), *The Social Impact of the Telephone*. Cambridge, MA: MIT Press.

Alker, H. 1966. Causal Inference and Political Analysis. *In* J. Beard (Ed.), *Mathematical Applications in Political Science II*. Dallas, TX: SMU Press.

American Telephone and Telegraph Company. Published annually. *The World's Telephones*. Bedminster, NJ: AT&T Long Lines Division, Overseas Administration.

Arellano Moreno, H. 1978. "Consideraciones Sociales en la Planificacion del Desarrollo de las Telecommunicaciones en las Regiones Distantes y Des favorecidas." Ottawa, Ontario, (June).*

Artle, R., and C. Averous. 1973. The Telephone System as a Public Good: Static and Dynamic Aspects. *Bell Journal of Economics 4*, pp. 89–100.

Astrain, S. 1978. "The Role of the INTELSAT System in Meeting International and National Telecommunications Needs." (Paper presented at International Satellite Communication Seminar, Lima, Peru.)

Bairi, A., and J. Leonhard. 1975. "A Domestic Satellite Communications System for Algeria." (Paper presented at IEEE Conference on Communications.)

Ball, D. W. 1968. Toward a Sociology of Telephones and Telephoners. *In* M. Truzzi (Ed.), *Sociology in Everyday Life*. Englewood Cliffs, NJ: Prentice-Hall.

Banks, A. 1971. *Cross-Polity Time-Series Data*. Cambridge, MA: MIT Press.

Baran, P. A. 1957. *The Political Economy of Growth*. New York: Monthly Review Press.

Beal, J. C. 1976. A Multidisciplinary Investigation of the Application of Dynamic Modeling to the Study of Telecommunications Development in Canada. *IEEE Transactions on Engineering Management 23* (2), pp. 89–92.

Bebee, E. L., and E. W. J. Gilling. 1976. Telecommunications and Economic Development: A Model for Planning and Policy Making. *Telecommunication Journal 43*, pp. 537–543.

Benstead, G. 1977. "USP Satellite Communication Project: Report for the Experimental Years 1975–76." Suva, Fiji: University of the South Pacific, Extension Services.

Berg, A. I. (Ed.). 1970. *Informacja i Cybernetyka*. (Information and Cybernetics). Warsaw. (Translated from Russian.)

Bernard, G. 1976. *Une Taxonomie d'Activités*. Paris: Cujas.

* Asterisked papers were presented at the Workshop on the Special Aspects of Telecommunications Development in Isolated and Underprivileged Areas of Countries, sponsored by the Department of Communications, Government of Canada, Ottawa, June 24–26, 1978.

Berry, J. F. 1981. "Comments on the Contribution of Telecommunications to Development with Particular Emphasis on France and Spain." Marnes-la-Coquette, France: Association Francais des Utilisateurs du Téléphone et des Télécommunications.

Berry J. F. 1977. Some Problems and Requirements of European Users of Telecommunications. *In* P. Polishuk and M. O'Bryant (Eds.), *Telecommunications and Economic Development.* (Papers presented at INTELCOM 77, Atlanta, GA) Dedham, MA: Horizon House International.

Blackman, R. C. 1977. "A Role for Telecommunications in the Economic Development of Africa and the Middle East." Unpublished thesis. University of Colorado, Department of Electrical Engineering.

Blackman, R. C. 1977a. Telecommunications for Economic Development in the Lower Income Countries. *In* P. Polishuk and M. O'Bryant (Eds.), *Telecommunications and Economic Development.* (Papers presented at INTELCOM 77, Atlanta, GA) Dedham, MA: Horizon House International.

Blanc, G. 1981. "The Impacts of Telecommunications on Employment." Paris: Development Centre, Organization for Economic Cooperation and Development.

Block, C. H. 1982. Promising Step at Acapulco: A U.S. View. *Journal of Communication 32* (3), pp. 60–70.

Bonilla, A. 1977. Rural Telecommunications in Costa Rica. *In* P. Polishuk and M. O'Bryant (Eds.), *Telecommunications and Economic Development.* (Papers presented at INTELCOM 77, Atlanta, GA) Dedham, MA: Horizon House International.

Bradford, D. E. 1971. Joint Products, Collective Goods, and External Effects: Comment. *Journal of Political Economy 79,* pp. 1119–1128.

Brunel, L. 1978. Télécommunications, des Machines et des Hommes. *Les dossiers de Québec Science.*

Buchanan, J. M. 1968. *The Demand and Supply of Public Goods.* Chicago, IL: Rand McNally.

Burrowes, R. 1970. Multiple Time-Series Analysis of Nation-Level Data. *Comparative Political Studies 2,* pp. 465–480.

Canas, M., and F. Antonio. 1977. System Development Patterns and Policies: The Case of Costa Rica." *In* P. Polishuk and M. O'Bryant (Eds.), *Telecommunications and Economic Development.* (Papers presented at INTELCOM 77, Atlanta, GA) Dedham, MA: Horizon House International.

Carreon, C. S. 1976. The Requirements of Developing Countries. *Telecommunications Journal 43,* pp. 124–129.

Chasia, H. 1976. Choice of Technology for Rural Telecommunications in Developing Countries. *IEEE Transactions on Communications 24,* pp. 732–736.

Cherry, C. 1977. The Telephone System: Creator of Mobility and Social Change. *In* I. de S. Pool (Ed.), *The Social Impact of the Telephone.* Cambridge, MA: MIT Press.

Cherry, C. 1971. *World Communication: Threat or Promise?* London: Wiley.

Clarke, D. G., and W. Laufenberg. 1981. "The Role of Telecommunications in Economic Development, with Special Reference to Rural Sub-Sahara Africa." Geneva: International Telecommunications Union.

Clippinger, J. H. 1977. Can Communications Development Benefit the Third World? *Telecommunications Policy 1,* pp. 295–304.

Criscolo, R. A. 1976. Rural Telecommunications in Latin America. *IEEE Transactions on Communications 24,* pp. 325–329.

Cruise O'Brien, R., and G. K. Helleiner. 1982. The Political Economy of Information in a Changing International Economic Order. *In* M. Jussawalla and D. M. Lamberton (Eds.), *Communication Economics and Development.* Elmsford, NY: Pergamon.

Cruise O'Brien, R., E. Cooper, B. Perkes, and H. Lucas. 1977. "Communications Indicators and Indicators of Socioeconomic Development, 1960–1970." Unpublished study. Sussex, England: University of Sussex, Institute of Development Studies.

Demmert, J. P., and J. L. Wilke. 1982. The Learn/Alaska Networks: Instructional Telecommunications in Alaska. *In Telecommunication in Alaska.* Honolulu, HI: Pacific Telecommunications Council.

Denison, E. 1974. Accounting for United States Growth 1929–1969. Washington, DC: Brookings Institution.

Denison, E. 1962. *The Sources of Economic Growth in the United States and the Alternatives Before Us.* New York: Committee for Economic Development.

Denison, E. 1967. *Why Growth Rates Differ: Postwar Experience in Nine Western Countries.* Washington, DC: Brookings Institution.

Dickenson, C. R. 1977. Telecommunications in the Developing Countries: The Relation to the Economy and the Society. *In* P. Polishuk and M. O'Bryant (Eds.), *Telecommunications and Economic Development.* (Papers presented at INTELCOM 77, Atlanta, GA) Dedham, MA: Horizon House International.

Dicks, D. 1977. From Dog Sled to Dial Telephone: A Cultural Gap? *Journal of Communication* 27(4), pp. 120–129.

Dormois, M., and M. Gensollen. 1976. Le marché du téléphone. *Economie et Statistique* (78), pp. 3–12.

Dubret, F. 1981. "Telecommunications and Their Impact on the Fishing Industry." Geneva: International Telecommunications Union.

Duncan, O. 1975. *Introduction to Structural Equation Models.* New York: Academic Press.

EDUTEL Communications and Development. 1978. "Computerized Telecommunications Planning Tools." (Mimeo.) Palo Alto, CA.

Encel, S. 1975. Social Aspects of Communication. *IEEE Transactions on Communication 23,* pp. 1012–1018.

Espinal, C. 1977. HONDUTEL: Our Case for National Telecommunications Planning. *In* P. Polishuk and M. O'Bryant (Eds.), *Telecommunications and Economic Development.* (Papers presented at INTELCOM 77, Atlanta, GA) Dedham, MA: Horizon House International.

Foote, D., E. B. Parker, and H. E. Hudson. 1976. "Telemedicine in Alaska: The ATS–6 Satellite Biomedical Demonstration." Stanford, CA: Institute for Communication Research, Stanford University.

Freire, P. 1970. *Pedagogy of the Oppressed.* New York: Seabury Press.

Freire, P. 1973. *Education for Critical Consciousness.* New York: Seabury Press.

Frey, F. A. 1973. Communication and Development. *In* I. de S. Pool, F. W. Fry, W. Schramm, N. Maccoby, and E. B. Parker. *Handbook of Communication.* Chicago, IL: Rand McNally.

Gellerman, R. F. 1977. Appropriate Technology in Latin American Telecommunications. *In* P. Polishuk and M. O'Bryant (Eds.), *Telecommunications and Economic Development.* (Papers presented at INTELCOM 77, Atlanta, GA) Dedham, MA: Horizon House International.

Gellerman, R. F. 1977a. Measuring the Benefits of Telecommunications. *In* P. Polishuk and M. O'Bryant (Eds.), *Telecommunications and Economic Development.* (Papers presented at INTELCOM 77, Atlanta, GA) Dedham, MA: Horizon House International.

Gellerman, R. F. 1978. "Telecommunications Activities of the Inter-American Development Bank." Washington, DC: Inter-American Development Bank.*

Gellerman, R. F., and S. Ling. 1976. Linking Electricity with the Telephone Demand Forecast: A Technical Note. *IEEE Transactions on Communications 24,* pp. 322–325.

Gilling, E. 1975. "Telecommunications and Economic Development: Inter-Country Comparisons of the Catalytic Effect of Telephone Services on Development." Unpublished MBA thesis, McGill University.

Gimpelson, L. A. 1974. Communications Planning for Developing Countries. *Telecommunication Journal 41,* pp. 486–491.

Gimpelson, L. A. 1976. Planning Communications Systems in Developing Countries. *IEEE Transactions on Communications 24,* pp. 710–715.

Giraud, A. 1978. *Les Réseaux Pensants, Télécommunications et Société.* Paris: Masson.

Goetschin, P. R. 1976. L'Importance des Télécommunications. *Bulletin de l'Association Suisse des Electriciens* (SEV/VSE). (May 15).

Golding, P. 1974. Media Role in National Development: Critique of a Theoretical Orthodoxy. *Journal of Communication 24*, pp. 39–53.

Goldschmidt, D. 1977. The Evolution of the Alaskan Satellite Telecommunications System. *In* P. Polishuk and M. O'Bryant (Eds.), *Telecommunications and Economic Development.* (Papers presented at INTELCOM 77, Atlanta, GA) Dedham, MA: Horizon House International.

Goldschmidt, D. 1978. "Telephone Communications, Collective Supply, and Public Goods: A Case Study of the Alaskan Telephone System." Unpublished dissertation, University of Pennsylvania.

Gorelik, M. A., and I. B. Efimova. 1977. The Economic Efficiency of Development of Long Distance Telephone Communication. *Vestnik Sviazi* (5). (In Russian.)

Gorelik, M. A., and E. Karaseva. 1975. Standards and Assessment of the Economic Efficiency of Long Distance Telephone Communication. *Vestnik Sviazi* (8). (In Russian.)

Gorelik, M. A., I. B. Efimova, and E. Karaseva. How to Determine the Economic Efficiency of Long Distance Telephone Communication. *Vestnik Sviazi* (7). (In Russian.)

Goulet, D. 1971. *The Cruel Choice.* New York: Atheneum.

Haas, M. 1966. Aggregate Analysis. *World Politics, 19,* pp. 106–121.

Hagen, E. 1975. *The Economics of Development.* Homewood, IL: Irwin.

Hannan, M. 1973. "Education and Economic Development: Some Baseline Models." Unpublished paper, Stanford University. Cited in J. Delacroix, "Information Processes and Economic Development: A Longitudinal Study of Dominance-Dependence Relationships in the World Ecosystem." Unpublished dissertation, Stanford University.

Hardy, A. P. 1980. The Role of the Telephone in Economic Development. *Telecommunications Policy 4,* pp. 278–286.

Hardy, A. P. 1981. "The Role of the Telephone in Economic Development: An Empirical Analysis." Geneva: International Telecommunications Union.

Harrois-Monin, F. 1976. Cure de Jouvence et Remèdes de Cheval pour un Téléphone Centenaire. *Science et Vie 701* (Feb.).

Havens, E. 1972. Methodological Issues in the Study of Development. *Sociologia Ruralis 12.*

Hoffman, E. 1975. Extended Evaluation of the Real Benefits Obtained Through Telephone Service Utilization. *Biuletyn Techniczny Ministerstwa Lacznosci* (2). (In Polish.)

Hoffman, E. 1974. An Attempt to Evaluate Real Benefits Obtained Due to Telephone Service Utilization. *Biuletyn Techniczny Ministerstwa Lacznosci* (3). (In Polish.)

Horley, A. L. 1976. "Broadband Electronic Communications for Brazil: An Investment Decision Analysis." Unpublished dissertation, Stanford University.

Hornik, R. 1980. Communication as Complement in Development. *Journal of Communication 30* (2), pp. 10–24.

Hudson, H. E. 1974. "Community Communications and Development: A Canadian Case Study." Unpublished dissertation, Stanford University.

Hudson, H. E. 1977. The Role of Radio in the Canadian North. *Journal of Communication 27* (4), pp. 130–139.

Hudson, H. E. 1981. "Three Case Studies on the Benefits of Telecommunications in Socio-Economic Development: The Cook Islands Interisland Radio System, The University of the South Pacific Satellite Network, Satellite-Delivered Telecommunications Services in Alaska." Geneva: International Telecommunications Union.

Hudson, H. E., and E. B. Parker. 1973. Medical Communication in Alaska by Satellite. *New England Journal of Medicine 289,* (Dec. 20), pp. 1351–1356.

Hudson, H. E., and E. B. Parker. 1975. Telecommunications: Planning for Rural Development. *IEEE Transactions on Communication 23,* pp. 1177–1185.

Hudson, H. E., D. Goldschmidt, E. B. Parker, and A. P. Hardy. 1979. "The Role of Telecommunications in Socio-Economic Development: A Review of the Literature with Guidelines for Further Investigations." Geneva: International Telecommunications Union.

Hudson, H. E., A. P. Hardy, and E. B. Parker. 1981. "Projections of the Installation of Telephones and Thin Route Satellite Earth Stations on Rural Development." Geneva: International Telecommunications Union.

Hudson, H. E., A. P. Hardy, and E. B. Parker. 1982. Impact of Telephone and Satellite Earth Station Installations on GDP. *Telecommunications Policy 6*, pp. 300–307.

Husarski, L. 1976. Efektywność Miedzymiastowego Ruchu Telefonicznego w Gospodarce Narodowej Zwiazku Radzieckiego. (Efficiency of Long Distance Telephone Service in the Soviet national Economy). *Wiadomosci Telekomunikacyjne 6*.

IDB News. 1977. Telecommunications in Latin America: Still More to Be Built. *Inter-American Development Bank 4*(7). Washington, DC: Inter-American Development Bank.

Indian Posts and Telecommunications Board. 1978. "Survey of Public Call Offices." Delhi, India: National Council of Applied Economic Research.

Inkeles, A. 1969. Making Men Modern: On the Causes and Consequences of Change in Six Developing Countries. *American Journal of Sociology 75*, pp. 208–225.

Innis, H. 1950. *Empire and Communication*. London: Clarendon Press.

Innis, H. 1951. *The Bias of Communication*. Toronto, Ontario: University of Toronto Press.

International Bank for Reconstruction and Development, Public Utilities Department. 1974. "Economic Evaluation of Public Utility Projects." (Mimeo.)

International Telecommunications Union. 1972. "Seminar on the Planning and Development of Telecommunications Networks Outside of Large Cities and the Maintenance of Telecommunications Services." Geneva: International Telecommunications Union.

International Telecommunications Union. 1982. *Yearbook of Common Carrier Telecommunications Statistics and Radio Communications Statistics*. Geneva: International Telecommunications Union.

International Telegraph and Telephone Consultative Committee. (CCITT). Autonomous Study Group (GAS) 5. 1968, 1972, 1976. *Economic Studies at the National Level in the Field of Telecommunications*. (GAS 5 Manuals.) Geneva: International Telecommunications Union.

Inuit Tapirisat of Canada. 1978. "Intervention before the Canadian Radio-Television and Telecommunications Commission." Ottawa: CRTC.

Izzo, L. L. C. 1977. Rural Telephony in the State of Sao Paolo, Brazil. *In* P. Polishuk and M. O'Bryant (Eds.), *Telecommunications and Economic Development*. (Papers presented at INTELCOM 77, Atlanta, GA) Dedham, MA: Horizon House International.

Janky, J., and J. Barewald. 1977. "Interference Control in Broadcast Satellite Applications: Antenna Sidelobe Patterns and Transponder Transfer Gain." Palo Alto, CA: EDUTEL Communications and Development.

Jannès, H. 1970. *Le Dossier Secret du Téléphone*. Paris: Flammarion.

Jeancharles, M., J. -L. Popovics, J. Royer, and J. Verdier. 1981. "Examen de Quelques Solutions Techniques au Service des Télécommunications Rurales." Paris: CIT Alcatel.

Jéquier, N. 1976. *La Technologie Appropriée*. Paris: Centre de Développement de l'OCDE.

Jéquier, N. 1983. "The Indirect Employment-Generative Effects of Investments in Telecommunications." (Paper presented at World Communication Year Seminars on Telecommunications for Development).

Johnston, J. 1972. *Econometric Methods*. New York: McGraw-Hill.

Jussawalla, M. 1982. International Trade Theory and Communications. *In* M. Jussawalla and D. M. Lamberton (Eds.), *Communication Economics and Development*. Elmsford, NY: Pergamon.

Kamal, A. A. 1981. "A Cost Benefit Analysis of Rural Telephone Service in Egypt." Cairo: Cairo University.

Karunaratne, N. E. 1982. Telecommunications and Information in Development Planning Strategy. *In* M. Jussawalla and D. M. Lamberton (Eds.), *Communication Economics and Development.* Elmsford, NY: Pergamon.

Kaul, S. N. 1978. "The Integration of Telecommunications Services Planning with Planning in Other Sectors." New Delhi: Indian Posts and Telegraphs.

Kaul, S. N. 1981. "India's Rural Telephone Network." New Delhi: Economics Study Cell, Posts and Telegraphs Boards, Ministry of Communications of India.

Kenny, D. A. 1975. Cross-lagged Panel Correlation: A Test for Spuriousness. *Psychological Bulletin 82,* pp. 887–903.

Kerlinger, F., and E. Pedhazur. 1973. *Multiple Regression Analysis in Behavioral Research.* New York: Holt, Rinehart and Winston.

Kochen, M., and K. W. Deutsch. 1969. Toward a Rational Theory of Decentralization: Some Implications of a Mathematical Approach. *American Political Science Review 63,* pp. 734–749.

Kourourma, M. Y. 1978. "Considerations Sociales sur la Planification du Développment des Télécommunications dans les Régions Isolées et Défavorisées des Pays." Republic of Upper Volta: Office des Postes et Télécommunications.

Kreimer, O. 1975. "Telecommunications in Medicine: Interactive Satellite Radio for Health Care in Village Alaska." Unpublished dissertation, Stanford University, CA.

Kurland, J., P. Sawitz, and G. Osborne. 1977. "Assessment of Candidate Communications Satellite for Developing Countries." (Final report to the Agency for International Development.) Silver Spring, MD: ORI.

Laffay, J. 1968. *Les Télécommunications (Que Sais-je?, 335).* Paris: PUF.

Lerner, D. 1958. *The Passing of Traditional Society.* Glencoe, IL: Free Press.

Lesser, B. 1978. "Methodological Issues Involved in Assessing the Economic Impact of Telecommunications with Special Reference to Isolated and Underprivileged Areas." Halifax, Nova Scotia: Dalhousie University.*

Lesser, B., and L. Osberg. 1978. "The Importance of Telecommunications to Regional Economic Development, Phase I Report." Halifax, Nova Scotia: Government Studies Programme, Dalhousie University.

Lesser, B., and L. Osberg. 1981. "The Socio-Economic Development Benefits of Telecommunications." Geneva: International Telecommunications Union.

Levy, M., Jr. 1966. *Modernization and the Structure of Societies.* Princeton, NJ: Princeton University Press.

Libois, L. J. 1976. Vers une Approche Globale du Problème des Téléccommunications.: *Revue des PTT.*

Lonnstrom, S., and I. Moo. 1977. Economic Aspects in Telephone Systems. *In* P. Polishuk and M. O'Bryant (Eds.), *Telecommunications and Economic Development.* (Papers presented at INTELCOM 77, Atlanta, GA) Dedham, MA: Horizon House International.

Maddick, H. 1963. *Democracy, Decentralization, and Development.* New York: Asia Publishing House.

Many Voices, One World: Report by the International Commission for the Study of Communication Problems (MacBride Commission). 1980. Paris: UNESCO.

Marsh, D. 1976. Telecommunications as a factor in the economic development of a country. *IEEE Transactions on Communications 24,* pp. 716–722.

McAnany, E. G. (Ed.). 1980. *Communications in the Rural Third World: The Role of Information in Development.* New York: Praeger.

McCrone, D. J., and C. F. Cnudde. 1967. Toward a Communications Theory of Democratic Political Development: A Causal Model. *American Political Science Review 61,* pp. 72–79.

Mednikov, D. 1975. Kakoj Effekt daet Vnedrenie Dispetcerskoj Svjazi v Sel'chochozjajstvennoe Proizvodstvo? (What is the Effect of Dispatcher Communication in Farming?) *Vestnik Svjazi 4.*

Meier, R. 1962. *A Communications Theory of Urban Growth.* Cambridge, MA: MIT Press.

Melody, W. H. 1978. "Telecommunications in Alaska: Economics and Public Policy." Anchorage AK. Governor's Office of Telecommunications.

Mgaya, F. M. 1978. "Economic Considerations in Planning Telecommunications Development in Isolated and Underprivileged Areas." Tanzania Posts and Telecommunications Corporation.*

Mishan, E. J. 1969. The Relationship Between Joint Products, Collective Goods, and External Effects. *Journal of Political Economy 77,* pp. 329–348.

Mishan, E. J. 1971. The Postwar Literature on Externalities: An Interpretive Essay. *Journal of Economic Literature 9,* pp. 1–28.

Mitchell, W. C. 1975. "The Use of Satellites in Meeting the Telecommunications Needs of Developing Countries." Stanford, CA: Stanford University, Communication Satellite Planning Center. (Technical Report Number 1.)

Montmaneix, M. G. 1974. *Le Téléphone. (Que Sais-je?, 251).* Paris: PUF.

Myrdal, G. 1970. *The Challenge of World Poverty.* New York: Penguin.

National Council of Applied Economic Research (India). 1978. "Survey of Rural Public Call Offices." New Delhi: Posts and Telegraphs.

Nickelson, R. L., and G. F. Tustison. 1981. An Earth Station Design for Rural Telecommunications. (Mimeo.) Geneva: International Telecommunications Union.

Nie, N., C. Hull, J. Jenkins, K. Steinbrenner, and D. Bent. 1975. *Statistical Package for the Social Sciences,* 2nd ed. New York: McGraw-Hill.

Nora, S., and A. Minc. 1980. *The Computerization of Society.* Cambridge, MA: MIT Press.

North–South: A Program for Survival: Report of the Independent Commission on International Development Issues (Brandt Commission). 1980. London: Pan Books.

Office of Telecommunications Policy. 1977. "Interagency Report on Rural Communications." Washington, DC: Executive Office of the President, (Dec. 15).

Ohlman, H. 1981. "Telecommunication Transport Tradeoffs." Geneva: International Telecommunications Union.

Okundi, P. O. 1978. "The Integration of Telecommunication Services Planning with Planning in Other Sectors." Nairobi, Kenya: Kenya Posts & Telecommunications Corporation.*

Okundi, P. O. 1976. Pan-African Telecommunications Network: A Case for Telecommunications in the Development of Africa. *IEEE Transactions on Communications 24,* pp. 749–755.

Okundi, P. O., A. W. Ogawo, and J. P. Kibombo. 1977. Rural Telecommunications Development in East Africa. *In* P. Polishuk and M. O'Bryant (Eds.), *Telecommunications and Economic Development.* (Papers presented at INTELCOM 77, Atlanta, GA) Dedham, MA: Horizon House International.

Osborne, W. P., R. L. Smith, and H. J. Stapon. 1977. SUDOSAT: The National Domestic Satellite Communications Systems for the Government of the Democratic Republic of the Sudan. *In* P. Polishuk and M. O'Bryant (Eds.), *Telecommunications and Economic Development.* (Papers presented at INTELCOM 77, Atlanta, GA) Dedham, MA: Horizon House International.

Parker, E. B. 1976. "Planning Communication Technologies and Institutions for Development." (Paper for East–West Center Conference on Communication Policy and Planning for Development, April.)

Parker, E. B. May 1978a. "Communication Satellites for Rural Development." (Paper presented to the International Satellite Communication Seminar, Lima, Peru.)

Parker, E. B. 1978b. An Information-Based Hypothesis. *Journal of Communication 28* (1), pp. 81–83.

Parker, E. B. 1981. "Economic and Social Benefits of the REA Telephone Loan Program." Mountainview, CA: Equatorial Communications.

Parker, W. B. 1973. "Village Satellite II—The Second Evaluation of Some Educational Uses of the ATS-1 Satellite Educational Broadcasting in Alaska." Anchorage, AK.

Pelz, D. C., and F. M. Andrews. 1964. Detecting Causal Priorities in Panel Study Data. *American Sociological Review 29*, pp. 836–848.

Phuc, N. T., and G. Dennery. 1972. *L'Économie des Télécommunications*. (Collection SUP). Paris: PUF.

Pierce, W., and N. Jéquier. 1977. The Contribution of Telecommunications to Economic Development. *Telecommunication Journal 44*, pp. 532–534.

Pierce, W. and N. Jéquier, 1983. "Telecommunications for Development: Synthesis Report of the ITU-OECD Project on the Contribution of Telecommunications to Economic and Social Development." Geneva: International Telecommunications Union.

Pierson, J. H. S. 1977. "The Economic Need for Telecommunications in Developing Nations." (Paper for the Joint ORSA/TIMS Meeting, Atlanta, GA, Nov. 8.)

Pinheiro, R. G. 1976. "Investment and Traffic Assignment in a Hybrid Ground and Spatial Communication Network." Stanford, CA: Stanford University, Communication Satellite Planning Center. (Technical Report Number 4.)

Pitroda, S. D. 1976. Telecommunication Development—The Third Way. *IEEE Transactions on Communication 24*, pp. 736–742.

Polishuk, P., and M. O'Bryant. (Eds.). 1977. "Telecommunications and Economic Development." (Papers Presented at INTELCOM 77, Atlanta, GA) Dedham, MA: Horizon House International.

Pool, I. de S. (Ed.). 1977. *The Social Impact of the Telephone*. Cambridge, MA: MIT Press.

Pool, I. de S., E. Freedman, and C. Warren. 1976. *Low Cost Data Communication for the Less Developed Countries*. Cambridge, MA: MIT.

Prasada, B. 1976. Some Implications of Telecommunication Policies in Developing Countries. *IEEE Transactions on Communications 24*, pp. 729–732.

Prentis, M. R. 1978. "Some Economic Considerations in Planning Telecommunications Services Development in Isolated and Underprivileged Areas of Canada." Ottawa, Ontario: Department of Communications.*

Raman, S. S. 1978. "Economic Considerations in Telecommunications Develoment in Isolated and Backward Areas." Delhi, India: Posts and Telegraphs Board.*

"Rapport de la Commission Transports et Communications, Préparation du VIIème Plan." 1976. Paris: Commissariat Général du Plan, Documentation Française.

Research Institute for Telecommunications and Economics. 1978. (Translated excerpts from a report on telecommunication's role in economic development.) Tokyo, Japan.

Rizzoni, E. M. 1977. Worldwide Telecommunications: System Development Patterns and Policies. *In* P. Polishuk and M. O'Bryant (Eds.), *Telecommunications and Economic Development*. (Papers presented at INTELCOM 77, Atlanta, GA) Dedham, MA: Horizon House International.

Rizzoni, E. M. 1976. An Overview of Latin American Telecommunications, Past, Present, and Future. *IEEE Transactions on Communications 24*, pp. 290–305.

Rogers, E. M. 1976. Communication and Development: The Passing of the Dominant Paradigm. *Communication Research 3*, pp. 213–240.

Rogers, E. M. 1972a. The Communication of Innovations: Combining Mass Media and Interpersonal Channels. *In* R. Solo and E. M. Rogers, (Eds.), *Inducing Technological Change for Economic Growth and Development*. East Lansing, MI: Michigan State University Press.

Rogers, E. M. 1972b. Key Concepts and Models. *In* R. Solo and E. M. Rogers. (Eds.), *Inducing Technological Change for Economic Growth and Development*. East Lansing, MI: Michigan State University Press.

Rogers, E. M., and F. F. Shoemaker. 1971. *Communication of Innovations*. New York: Free Press.

Russell, S. 1977. "Techno-Economics of U.S. Domestic Satellite Orbit-Spectrum Utilization." Unpublished dissertation, Stanford University, CA.

Russett, B., (Ed.). 1964. *World Handbook of Political and Social Indicators*. New Haven, CT: Yale University Press.

Saunders, R. 1982. "Telecommunications in Developing Countries: Constraints on Development." *In* M. Jussawalla and D. M. Lamberton (Eds.), *Communication Economics and Development.* Elmsford, NY: Pergamon.

Saunders, R. 1978. "Rural Telecommunications: Economic and Policy Implications." (Paper delivered at the ITU Seminar on Rural Telecommunications, New Delhi, Aug.)

Saunders, R., and J. Warford. 1977. Telecommunications Pricing Investment in Developing Countries. *In* P. Polishuk and M. O'Bryant (Eds.) *Telecommunications and Economic Development.* (Papers presented at INTELCOM 77, Atlanta, GA) Dedham, MA: Horizon House International.

Saunders, R., and J. Warford. 1978. "Evaluation of Telephone Projects in Less Developed Countries." Washington, DC: World Bank. (P.U. Report No. PUN 37.)

Saunders, R., J. Warford, and B. Wellenius. 1983. *Telecommunications and Economic Development.* Baltimore, MD: Johns Hopkins University Press.

Scheele, C. H. 1970. *A Short History of the Mail Service.* Washington, DC: Smithsonian Institution.

Schramm, W. 1964. *Mass Media and National Development.* Stanford, CA: Stanford University Press.

Schramm, W., and W. W. Ruggles. 1967. How Mass Media Systems Grow. *In* D. Lerner and W. Schramm (Eds.), *Communications and Change in the Developing Countries.* Honolulu, HI: East-West Center Press.

Sesay, S. J. 1977. Some Problems of Improving Telecommunications Systems in the Developing Countries. *In* P. Polishuk and M. O'Bryant (Eds.), *Telecommunications and Economic Development.* (Papers presented at INTELCOM 77, Atlanta, GA) Dedham, MA: Horizon House International.

Shapiro, P. D. 1976. Telecommunications and Industrial Development. *IEEE Transactions on Communications 24,* pp. 305–311.

Shapiro, P. S. 1967. "Communication or Transport: Decision-Making in Developing Countries." Cambridge, MA: Center for International Studies, MIT.

Sharma, R. G. 1976. "Rural Communications Planning Methodology for Integrating Satellite and Terrestrial Facilities." Stanford, CA: Stanford University, Communication Satellite Planning Center. (Technical Report Number 66)

Simpson, A. A. 1978. "Remote Communication in Canada and Social Interaction—A Middle Management Perspective." Ottawa, Ontario: Department of Communications.*

Sites, M. 1976. "A Demand Assignment Multiple Access (DAMA) Control System for Thin-Route Satellite Telephone Systems." Stanford, CA: Stanford University, Communication Satellite Planning Center. (Technical Report Number 10.)

"Soviet Studies on the Indirect Benefits of Investment in Telecommunications." 1978. (Mimeo). Paris: OECD Development Centre.

"Space Radiocommunications System for Aid Following Natural Disasters." 1975. (ITU Booklet No. 18.) Geneva: International Telecommunications Union.

Tanhuanpaa, A. R. 1974. The Socio-Economic Importance of Teletraffic Development. *Sahko: Electricity in Finland 47* (5–6)

Taylor, C. L., and M. C. Hudson. *World Handbook of Political and Social Indicators.* (2nd ed.) New Haven, CT: Yale University Press.

Teer, K. 1975. Communication and Telecommunication. *IEEE Transactions on Communication 23,* pp. 1040–1045.

Thorngren, B. 1977. Silent Actors: Communications Networks for Development. *In* I. de S. Pool (Ed.), *The Social Impact of the Telephone.* Cambridge, MA: MIT Press.

Traubenberg, I. A., and Z. E. Patrunova. 1973. O Vlijanii Urovnja Razvitija Sredstv Svjazi Processy Upravlenija. (On the Influence of the Level of Development of the Means of Communication on the Process of Management.) *Trudy Ucebnych Institutov Sbjazi 65.* (In Russian.)

Tyler, M. 1981. "The Impact of Telecommunications on the Performance of a Sample of Business Enterprises in Kenya." Geneva: International Telecommunications Union.

Tyler, M., R. Morgan, and A. Clarke., 1981. "Telecommunications and Energy Policy." New York: Communications Studies and Planning International.

UNESCO. *Statistical Yearbook.* (published annually) Paris: UNESCO.

United Nations. *Statistical Yearbook.* (published annually) New York: United Nations.

Valerdi, J. 1977a. A Communications Plan for Mexico: Opportunity for Recovery. *Telecommunications Policy 1*, pp. 271–288.

Valerdi, J. 1977b. *Mexican telecommunications for the year 2000. In* P. Polishuk and M. O'Bryant (Eds.), "Telecommunications and Economic Development." (Papers presented at INTELCOM 77, Atlanta, GA) Dedham, MA: Horizon House International.

Velasquez, A., and D. Angel. "Planning the Optimum Utilization of Preassigned and Demand Assignment Services in Domestic Satellite Telecommunication Systems." Stanford, CA: Stanford University, Communication Satellite Planning Center. (Technical Report Number 5.)

Von Rabenau, B., and K. Stahl. 1974. Dynamic Aspects of Public Goods: A Further Analysis of the Telephone System. *Bell Journal of Economics 5*, pp. 651–669.

Voronov, B. A. 1969. The Consumption of Communications Output. *Vestnik Sviazi* (8). (In Russian.)

Walp, R. M. 1977. "The Evolution, Design and Use of a Telecommunications System in a Lightly Populated Region." Juneau, AK: Governor's Office of Telecommunications.

Wardrop, D. H. 1978. "Telephone Assistance Plan." Winnipeg, Manitoba: Manitoba Telephone System.*

Wa-Wa-Ta Native Communications Society. 1978. "Intervention Before the Canadian Radio-Television and Telecommunications Commission." Ottawa: CRTC.

Wcisly, K. 1978a. "The Role of Telecommunications in Socio-Economic Development: A Review of Empirical Studies in Poland and the USSR." (Mimeo.) Paris: OECD Development Centre.

Wcisly, K. 1978b. Telecommunications—Development Factor. *Przaglad Techniczny 10*. (In Polish.)

Wcisly, K. 1977a. A New Generation of Management Systems. *Zycie Warszawy 271*. (In Polish.)

Wcisly, K. 1977b. Do You Have a Telephone? *Zycie Gospodarcze 40*. (In Polish.)

Wcisly, K. 1976. The Role of Telecommunications in Economic Growth. *Wybrane Aktualne Problemy Telekomunikacju.* Warsaw: Instytut Zacznosci. (In Polish.)

Weldon, K. L., J. Schulman, and B. Lesser. 1977. "Economic Benefits of Improved Telephone Service to Rural Areas: Phase I Report." Unpublished report. Halifax, Nova Scotia: Dalhousie University.

Wellenius, B. 1971. Estimating Telephone Necessities in Developing Nations. *IEEE Transactions on Communication Technology 19*, pp. 333–339.

Wellenius, B. 1972. *On the Role of Telecommunications in the Development of Nations. IEEE Transactions on Communication Technology 20*, pp. 3–8.

Wellenius, B. 1978. "The Role of Telecommunication Services in Developing Countries." Universidad de Chile, Departmento de Electricidad.*

Wellenius, B. 1977. Rural Telecommunications in the Developing Countries: Introduction." *In* P. Polishuk and M. O'Bryant (Eds.) *Telecommunications and Economic Development.* (Papers presented at INTELCOM 77, Atlanta, GA) Dedham, MA: Horizon House International.

Wellenius, B. 1977. Telecommunications in Developing Countries. *Telecommunications Policy 1*, pp. 289–297.

Wellenius, B., L. Castillo, and E. Melnick. 1971. "Methodologia para el Estudio de las Telecomunicaciones Rurales: Cuatro Contribuciones." Department of Electrical Engineering, University of Chile. (Research Report 28.)

White, C. E. 1976. Telecommunications Needs of Developing Countries. *Telecommunications 10*(9), pp. 49–63.

Whyte, J. S. 1971. "Telecommunications in the Service of Man." ICC Conference paper.

Winham, G. R. 1970. Political Development and Lerner's Theory: Further Test of a Causal Model. *American Political Science Review 64*, pp. 810–818.

Wolter, W. 1977. Improved Telecommunications—Key to World Economic Growth. *Telephony 192*(20), pp. 86–88.

"World Administrative Radio Conference of 1979: Final Acts." 1979. Geneva: International Telecommunications Union.

World Bank. 1983. *World Development Report.* Washington, DC: World Bank.

World Bank. 1976. *World Tables 1976.* Baltimore, MD: John Hopkins University Press.

World Bank. 1971. "Telecommunications Sector Working Paper." (Mimeo.) Washington, DC: World Bank.

Wurtzel, A. H., and C. Turner. 1977. Latent Functions of the Telephone: What Missing the Extension Means. *In* I. de S. Pool (Ed.), *The Social Impact of the Telephone.* Cambridge, MA: MIT Press.

Author Index

Italic page numbers indicate bibliographic citations.

Subject Index

A

Access
  index of, 104–106
  to telephones, 129–130
Africa, 69–71
After-only designs, 109–110
Aggregate data analysis, 114–117
Agricultural marketing, 67–69
Agriculture, 58, 58–61, 61
Alaska, 30–31, 76–79
Alaska Native Land Claims Settlement Act of 1971, 30–31
Anik-B satellite, 9
Applied Technology Satellites (ATS), 9
Atka, 30

B

Benefits
  constraints limiting, 126–127
  factors related to, 130
  unresolved issues related to, 131–132
Bivariate relationships, 114–115
Brandt Commission, 11
Broadcasting, 13, 71
Business function of telecommunications, 81–83
Business telephones, 42–43

C

Calling externalities, 22

Canada, 30
Case studies
  as research approach, 17, 108–109
  value and limitations of, 101–102
Catalytic effect, 40
Causality question, 41–46
Commerce in developing countries. *See under* Developing countries
Communication in developing countries, 90–92
Communication variables, 110
Communities of interest, 130–131
Complementarity, 19, 25–28, 107
Consistency, need for, 132
Constraints
  on impact of telecommunications, 18–19
  institutional, effect of, 132–133
  on planning, 126–127
Consultative Committees, 10
Cook Islands two-way radio network, 67–69
Correlational approach, 39–40
Correlations, 115, 116
Cost(s)
  -benefit analyses, 112
  as component of index of access, 104
  of default (coût de défaillance), 54
  of poor telecommunications (economic), 52–55
  private, 21
  of rural economic activities, 24
Cross-lagged correlations, 116